Para-Interactivity and the Appeal of Television in the Digital Age

Para-Interactivity and the Appeal of Television in the Digital Age

Oranit Klein Shagrir

LEXINGTON BOOKS
Lanham • Boulder • New York • London

Published by Lexington Books
An imprint of The Rowman & Littlefield Publishing Group, Inc.
4501 Forbes Boulevard, Suite 200, Lanham, Maryland 20706
www.rowman.com

Unit A, Whitacre Mews, 26-34 Stannary Street, London SE11 4AB

British Library Cataloguing in Publication Information Available

Library of Congress Cataloging-in-Publication Data Available

ISBN 978-1-4985-4080-3 (cloth : alk. paper)
ISBN 978-1-4985-4081-0 (electronic)

Printed in the United States of America

Contents

Acknowledgments

Language editing completed by Sara Kitai.

Cover photo edited by Elad Shagrir.

Material from chapter 5 was previously published as an article:
Klein Shagrir, O. (2015). Unveiling television's apparatus on screen as a "para-interactive" strategy. *Media, Culture & Society*, 37(5), 737–52.

Preface

At an international TV industry conference called Innovation TV (INTV) held in Jerusalem (March, 2017),[1] Jon Steinberg, CEO of Cheddar TV,[2] opened his talk with the provocative statement: "We are living in a post TV era. TV is over even if TV people deny it." Murmurs of disagreement ran through the audience. The presentations and panel discussions that followed in the course of the two-day conference proved that broadcast television might be changing, but it is still alive and kicking, and has a commercial role, as executives from around the world insisted. One speaker put it succinctly: "TV still has the ability to deliver the mass reach advertisers require."[3] That said, television has been pondering its future for some time now, and fears of the demise of linear broadcasting are voiced by many in the industry. TV makers invest a lot of thought and effort into trying to figure out how to appeal to their audience, and above all, how to continue to attract viewers, in order to remain profitable. As the conference participants testified, their efforts are directed toward a variety of endeavors, including the production of sophisticated scripted dramas, high profile documentaries, specialized reality shows, live sports broadcasts, and innovative formats that can be distributed internationally. Another widespread notion among the movers and shakers of the TV industry, and one which may help explain the rather frenzied activity we are seeing at the moment is that younger audiences' preferences in the digital age are changing, and they are now eager to play an active part in media. In the words of a TV executive at the conference: "A good story is not enough for this generation. They are looking for engagement, connectivity, sharing, taking part."[4]

This book focuses on broadcast TV's attempts to transition from a mass medium, characterized by supposedly "passive" viewing, to one which addresses viewers as potentially active participants in simulated interactive communication. It explores TV's efforts to remain culturally relevant, socially significant, and commercially viable in a changing media environment.

I have been studying TV's challenges and its struggle to contend with them for the past ten years. The desire to investigate how TV is coping in these stressful times, and indeed to write this book, stemmed not only from academic interest, but also from the concerns regarding broadcast TV's future that I repeatedly heard from industry people. Moreover, from a personal point of view, like many people in my generation, television played a huge role in my early life and in my case at least, later in my professional life as well.

THE ROLE OF BROADCAST TV: A PERSONAL PERSPECTIVE

My fascination with TV started more than forty years ago as a young girl, and has persisted ever since. It inspired my career as an adult, first as a producer in the Israeli TV industry and then as an academic researcher of television.

The year 1973 is remembered by many people in Israel as a year of war with our neighbors, Egypt and Syria. Although far from Israel's only war, it sent shockwaves through the nation both because of its unexpectedness and because of the horrifying number of casualties. For me it was also the year of television, or more accurately, the year I became fascinated by television. The war broke out on the holiest day in the Jewish calendar, Yom Kippur or the Day of Atonement, the only day in the whole year when radio and television completely cease all transmissions. On October 6, at 2:00 pm, the sirens sounded and broke the silence of the holy day, and I remember my parents' faces: baffled, scared, not knowing what was happening. Immediately, they switched on the radio. Television had started broadcasting in Israel only a few years before, and people still depended on the radio for essential and reliable information around the clock. A short while after the first siren pierced the air, the radio transmitted its first newscast of the day, announcing that war had broken out and that all reserve soldiers must report to their units immediately. I watched as my father put on his uniform, said good-bye and left to join his army unit. I did not know then that it would be several months until we saw him again. Even though I was just a little girl, this was not my first war. I still had vivid memories of the Six-Day War only a few years earlier in 1967. Still, the 1973 war was Israel's first televised war, and as I would soon discover, that made a big difference. TV had commenced broadcasting in Israel in 1968, and gradually became a source of information, entertainment, comfort, and national togetherness. During the long cold evenings of the winter of 1973, my mom and I would sit in front of the TV set and watch the terrifying coverage of the war: the battles, the funerals, the wounded in the hospitals, the Israeli soldiers taken prisoner. We sat silently, staring intently at the screen and scanning the fleeting frames, praying to get a glimpse of my dad. He was fighting at the northern frontier and we had hardly heard anything from him. As the number of dead and wounded rose, so did are our anxieties and the fears of many other families around the nation. The TV screen was the only source from which we had any hope of getting a sign of life from him. At the time, the TV reporters who accompanied the troops seemed to me the real heroes of the war. Thanks to them, we all felt as if we were actually witnessing what was going on at the front. Not only did they inform us of the progress of the war, but they also had the power to show the tired face of a son or a father sending regards to his worried family at home. TV was also the place where the leaders of the nation spoke to us, trying to reassure us even when things were not going well. Looking back, I realize that TV seemed almost magical to me at the time, providing us with great comfort while at the same time arousing petrifying fears. Hence, my overwhelming excitement when a friend of the family, who worked as a cameraman for what was then the only TV channel in the country, offered to take us on a private tour of the studio in order to boost our morale. The beginning of the tour was a huge disappointment. "TV" turned out to be a gray, rather run-down office building. We were taken to the various floors, where we saw offices and editing rooms with big editing tables. When we peeked into the news studio, I was surprised by how small it was. At that point, maybe sensing my disappointment, our guide asked if we would like to see "the clock." Any Israeli would know what he meant: the ticking on-screen clock that counted down the final minute until the beginning of the main evening newscast and was a central feature of the evening routine in almost every Israeli home. As soon as the clock came on screen, it was a signal to hurry up, leave whatever you were doing, and sit down in front of the TV to watch the newscast with the rest of the nation. Obviously, there was no question that I wanted to see it, so we climbed the stairs to the fifth floor and looked through a small window into a tiny room and there it was: a simple wood board with 3 hands sitting on an old chair with a big camera pointing at it. This was "the clock," the famous icon that set the pace of life for a whole nation. It was a long time ago, but I can still remember the amazement I felt. In hindsight I can say that at that moment, in a young girl's mind, I understood TV's power to create magic by transforming a plain wooden board into a "national clock" that played such a significant role in people's lives. The visit to the studio also allowed me to see with my own eyes the immense gap between what we see on screen and the reality behind the scenes. When I watched them at home, the newscasters seemed so formal and authoritative in their suits and ties, but when I saw them in the small and unimpressive studio, I discovered to my shock that some of them were wearing shorts and sandals, which were obviously hidden from the viewers.

That day I decided I wanted to work in television, and ten years later I was able to realize my dream. I was only twenty years old when I got a job as a production assistant, and felt I was the luckiest girl in the world. There was still only one local channel, and anyone who appeared on screen immediately achieved superstar status. The journalists, newscasters, and performers were recognized and treated almost like family members. After all, they appeared in every living room around the country.

My early memories of TV's central role in my life and the lives of many Israelis are a striking backdrop to its changing status and diminishing power in society and culture in recent years. When examining the challenges TV is facing today, it is important to remind ourselves of its significance not only as a medium, but also as a social institution. I agree with Lotz's observation that TV is a "central communicative and cultural force within society. Its centrality derived from its availability and ubiquity" (2007, p. 31). At the same time, however, we can not ignore changing viewing habits and the threats to the economic model of broadcast TV. My own belief is that there is room in the widening media environment for linear TV alongside streaming services, interactive online media, and social media. That being said, TV will undoubtedly continue to undergo changes and will have to find a new, more adaptive commercial model and more creative ways to appeal to its audience if it wants to survive. This book, born out of my concerns for the future of broadcast TV combined with my deep awareness of its central role in my life, explores one aspect of its adaptation to the digital era; how TV attempts to appeal to its audience by using televisual strategies to simulate interactivity and viewer participation.

STRUCTURE OF THE BOOK

Chapter 1, "Conceptual Framework: TV, Interactivity and Para-Interactivity," introduces a new concept for the study of television—para-interactivity. The chapter begins with the background for the development of para-interactive strategies and discusses the dramatic changes in the media environment and the challenges they pose for broadcast TV. These changes, such as the increasing accessibility of means of production and distribution of media messages enabling interactivity, participatory culture, and the blurring of lines between production and consumption, are forcing television—a linear, top-down medium—to look for ways to preserve its relevance in the new mediascape. The chapter explores the main conceptual point of departure for this book, the ubiquitous concept of interactivity. I explain why I chose interactivity as my starting point, and then offer a brief review of the various definitions of the concept, which help in understanding the different elements comprising para-interactivity. The chapter goes on to discuss the interface between interactivity and TV, and compares the affordances (Hutchby, 2001) of TV with those of digital media and the changes in media audiences' expectations in view of these differences in the mediascape. This is followed by an in-depth explanation of the concept of para-interactivity, including its link to the legacy of TV research, the various strategies encompassed by the concept, and its meaning and purpose. Last but not least, I describe the methodologies I employed in the research that led to this book, the corpus studied, and some of the limitations of the project.

Chapter 2, "'1.0 Producers in a 2.0 World': The Dual Discourse of TV Makers" presents the main findings derived from the interviews I conducted with TV makers from 2008 to 2016. The chapter begins with a short review of the Israeli broadcast television industry and its challenges. This is followed by a description of the dual professional discourse that emerged from the interviews regarding the role of television and the interviewees' own professional skills in the digital environment. TV makers also expressed an agreement that audiences today wish to be active and interactive when consuming media and this impacts their expectations from television as well. The dual professional discourse and perception of the viewers is the background against which para-interactive strategies have evolved. The chapter ends with an explanation of the methodological obstacles encountered in studying the television industry.

Each of the next three chapters is devoted to one of the three categories of para-interactive strategies on three dimensions: paths of interaction, the temporal dimension, and the spatial dimension.

Chapter 3, "Para-Interactive Paths: Inviting the Viewers to Interact," explains the nature of para-interactive paths and the viewing experience they are aimed to produce, and then links the various types of paths to elements in the definitions of interactivity. Para-interactive paths are meant to create the impression that the viewers can be part of the televisual text, participate in it, impact it, engage with it, and invite viewers to position themselves in various attitudes in relation to the screen, the televisual text, and they themselves. The different para-interactive paths are described and demonstrated by numerous examples and quotes taken from TV texts and para-texts accompanying TV programs.

Chapter 4, "'Live' TV as a Para-Interactive Temporal Strategy," opens with the intriguing claim that "live" is the future of TV. This claim is discussed in view of the historic importance of live TV and recent developments that make a variety of potential temporalities available to the audience by means of services such as VOD and video streaming. The chapter then offers a brief review of the concepts of "live" and "liveness" in relation to television, and sketches the trajectory of their history. This is followed by an explanation of the strategies and means employed by contemporary broadcast television to create the appearance of "live television" through "liveness indicators," even when the shows are taped, as well as through new modes of liveness: "second shift liveness" and "alternative liveness" produced on online platforms and social media. Finally, the chapter explores the new meanings of live TV in the contemporary context of the digital mediascape, concluding that live TV is contextualized as a temporal para-interactive strategy that helps television strengthen its connection with viewers while at the same time enhancing and validating the para-interactive paths discussed in the previous chapter.

Chapter 5, "Unveiling TV's Apparatus as Para-interactive Spatial Strategy" presents the third avenue of para-interactivity. The chapter opens with the broadcast of the 2009 Emmy Awards, in which the producers decided to unveil the TV control room and place it on stage for the viewers to observe. This example demonstrates a growing phenomenon whereby the TV production apparatus is made more visible to the viewers at home, and thus the line between back and front stage is blurred. The chapter describes the spatial para-interactive strategy as a symbolic bidirectional spatial model. In one direction, television cameras move back and invite the viewers into the production space; in the opposite direction, elements that traditionally take place backstage are brought forward. This bidirectionality reflects the promise of symmetry and reciprocity between viewers and producers that is commonly associated with interactivity. This strategy is linked to the professional discourse and producers' perception of their audience as wishing to be part of the production of the program. In this context, the spatial para-interactive strategy is aimed at producing a seemingly inclusive and reciprocal relationship with the viewers. The strategy is contextualized theoretically with the help of Meyrowitz's Spatial Turn (1985) and Goffman's (1959) perspective of back and front stage social spaces. The analysis further explores the question of whether TV is changing its own spatial self-perceptions in response to the changes in the media environment. The chapter ends with a discussion of the concept of "staged backstage" (MacCannell, 1973), which leads to the conclusion that the backstage is only partially unveiled on screen and is actually staged to promote the interests of the producers.

NOTES

1. For conference site and list of speakers, see: http://www.in-tv.net/ (accessed: March 12, 2017).
2. http://www.cheddar.com/ (accessed: April 4, 2017).
3. Harris Diamond, Chairman and CEO McCann World Group (March 7, 2017), INTV conference, Jerusalem.
4. Scott Reich, Senior Vice President, Programming Full Screen (March 3, 2017), INTV conference, Jerusalem.

Chapter One

Conceptual Framework

TV, Interactivity and Para-Interactivity

While broadcast television is still socially and culturally relevant, it now operates in an environment inhabited by a wealth of digital platforms and social media, all competing for consumers' time and attention. Many of these platforms enable users to become producers and distributers of media content themselves. This book investigates the strategies broadcast television is now employing to appeal to viewers in an effort to maintain its sociocultural significance, as well as its revenues, in this challenging environment. As TV executives attest: "The technology is changing so fast around us but the measuring [of viewers] did not catch up with it. I am always looking for a show that appeals to a large amount of people" (Jennifer Salke, President, NBC Entertainment, 7.3.17).

In essence, TV is now attempting to manage its transition from a mass medium, characterized by supposedly "passive" viewing, to one which addresses viewers as potentially active participants in simulated interactive communication, or what I conceptualize as *para-interactivity*. I use this term to refer to a cluster of strategies that contain features or elements that mimic interactivity. However, as a rule, when employed by and adapted to TV, they do not actually construct interactive communication.

This opening chapter lays out the conceptual and methodological framework of the book. Here I will explicate the concept of para-interactivity and the conceptual starting point of interactivity, and detail the ways in which the research underlying the book was conducted. But first it is essential that we understand the transforming mediascape in which TV is operating today and the challenges it is facing.

TV AND THE DIGITAL AGE

In recent decades, the media environment has been undergoing a series of fundamental changes, with "old" mass media meant for "passive" consumption being joined by numerous "new" media that afford interactive communication (McQuail, 2010). Technological developments, such as video cameras and editing software, have become cheaper and simpler to operate, enabling more people—those "formerly known as the audience," (Rosen, 2006)—to become producers of media messages. In addition, new media, including webcams, smartphones, and digital platforms (YouTube, Instagram, and other social media), have made not only the means of content production, but also the means of distribution, widely accessible. This situation has blurred the line between the traditional roles of producers and receivers of media messages, and as a result, the boundaries between the traditional spheres of production and consumption of media have become blurred as well (Andrejevic, 2004a; Rice, 1999; Roig et al., 2009). Jenkins (2006), who focuses on technological convergence, conceives of the cultural practices that flourish around new digital media as "participatory culture." In other words, audience members are not only consumers of media, but also participants in it. Bruns (2006) refers to technology's potential to transform ordinary audience members into "producers" and focuses on community collaboration, claiming that users are becoming producers and vice versa, and that collaboration between participants is based on the principle of inclusivity. It is against the background of these changes in the media environment that I examine the ways in which television, a linear top-down medium, is striving to preserve its relevance.

The death of broadcast TV has been predicted since the mid-1990s. Some believed that old "passive" media would become obsolete and be replaced by the new interactive media, that is, the Internet would supplant broadcast TV (Jenkins, 2006). Michael Crichton, author of *Jurassic Park* (1990), compared old mass media to dinosaurs: huge, heavy, and slow-moving, it was incapable of adapting to the new environment and was therefore on the verge of extinction (cited in Jensen and Toscan, 1999). Nevertheless, broadcast TV is not yet extinct. On the contrary, over the years it has undergone constant changes in production, content, distribution, and consumption (Caldwell, 1995; Urichhio, 2002; Williams, 1974), and is continuing to do so. Recent technological developments have resulted in the transformation of TV from "a box" to a myriad of platforms and devices, and thus also to multiple reception contexts, as described in a rich body of research from the past decade (Katz and Scannell, 2009; Olsson and Spiegel, 2004; Turner and Tay, 2009). Some writers claim that in the wake of this transformation, the reference to old and new media is no longer warranted, since what we have now are simply various platforms and screens, and content can be consumed on any of them (Boyle, 2014). They argue that this is the context in which TV content and the industry that produces it should be studied (Sørensen, 2016a).

This book adopts an integrative approach. While it takes into account the fact that TV content is now spread across media and screens, it also investigates broadcast television as a "top-down" medium trying to adapt to a new reality in which it is forced to compete with interactive media. Even though TV content can be consumed in various venues, broadcast TV's fear of losing its audience to other platforms can not be ignored, particularly as its commercial model is still based primarily on counting eyeballs watching a linear transmission. Moreover, the data shows that linear TV's audience is aging thus as of 2016 the median age for CNN viewers was 61 and for MSNBC it was 63. Also in 2016 younger viewers spent 48 percent less time watching traditional TV in the U.S. than the time they spent watching in 2010.[1]

The conceptual point of departure for the analysis in this book is the concept of interactivity, with its many definitions. A central theme of both academic and popular discourse, interactivity is considered one of the main characteristics of new media, in contrast to older top-down media such as television (McQuail, 2010). Indeed, it is at the heart of digital media's promise to transform the audience into users and producers of media content, associated both with the changing role of audiences as depicted by the "participatory culture," and with the transformation of the relationship between producers and consumers of media (Jenkins, 2006). Moreover, interactivity is identified with enjoyment, pleasure and satisfaction and triggers feelings of "coolness, fun and playfulness" (Yang and Shen, 2017, p. 3) and therefore coveted by users and media producers alike.

The concept of interactivity has been defined in numerous ways, with each definition relating to a different set of elements (Kiousis, 2002). While this might seem to be a disadvantage for researchers, it proved to be an advantage for my investigation. It served me as an umbrella concept, with the various components helping to identify and chart the range of different strategies employed by television in an effort to adapt to the new interactive environment.

WHAT IS INTERACTIVITY?

Despite the prevalence of its use, the term *interactivity* still lacks a clear-cut definition (Kiousis, 2002). Rafaeli, one of the first to research the concept, simply claims that once you experience interactivity you recognize it (1988, p. 111). Others have compared the efforts to define it to the quest for the Holy Grail (Van Dijk and De Vos, 2001). As a result, there is ongoing debate in the literature over the definition and measurement of interactivity (Avidar, 2013). Scholars generally adopt one of three main perspectives: interactivity as a communication process; as a characteristic of a medium or technology; or as a perceived variable.

Researchers studying interactivity as a process of communication focus on message transmission, the reciprocity of the communication, the responsiveness of the participants, and the degree of interchange between them (Kelleher 2009; McMillan and Hwang 2002). Rafaeli (1988) formulated a highly specific definition of interactivity as "an expression of the extent that in a given series of communication exchanges, any third (or later) transmission (or message) is related to the degree to which previous exchanges referred to even earlier transmissions" (p. 111). He therefore distinguishes the interactive process from the more common reactive communication process, which involves only a message and a response to it. Another definition of this sort relates to the control the participants have over the communicative process and their ability to exchange roles in the reciprocal discourse (Williams et al., 1988, p. 111).

Those who view interactivity as the characteristic of a medium consider the affordances of the medium that allow users to participate in and impact the design, shape, and content of the mediatized environment in real time (Jensen, 1988b; Steuer, 1992). Thus, an interactive technological environment, such as a virtual environment, should afford responsiveness to the users' input (Ryan, 1994, p. 8). This approach tends to assign the labels of "low" or "high" interactivity to various media according to their technological features, so that high interactivity is ascribed to new media (such as smartphones and tablets) and low interactivity to traditional media (such as TV and radio). Research into interactivity as the characteristic of a medium has mainly been conducted in the context of new media or human—computer interaction (Janlert and Stolterman, 2017). However, it has also been noted that "the process of convergence links media that cannot enable interactivity with media that can, thus making it interactive" (Ariel and Avidar, 2015, p. 24).

The third approach to the definition of interactivity is to view it as a variable perceived by the users, a perspective adopted mainly in advertising and marketing research (McMillan and Hwang, 2002). Downes and McMillan (2000), for example, identified several recurring elements in the users' perceptions of interactivity: bi/multidirectional communication that allows all participants to be active; flexible or synchronic timing of the communication; the media environment produces a sense of place; the participants feel that they have control over the media environment; and the users perceive the medium as responsive to their input. Similarly, according to Ariel and Avidar (2015): "Popular conceptualizations of interactivity include synchronicity, control, rapidity and speed, participation, a variety of choices, directionality, hypertextuality, connectedness, experience, and responsiveness" (p. 21)

To sum up, despite the different approaches and multiple definitions, there appear to be several recurring basic components of interactivity: messages relate to some extent to previous messages; users' have a degree of control over the media environment and media content; producers and consumers of media are able to exchange roles; users are offered a range of choices; and the interaction is instantaneous and often synchronous. These features served as the foundation for my examination of TV's efforts to appeal to its audience in a digital and interactive environment. Moreover, they inspired key elements in my conceptualization of the central concept in this book: para-interactivity.

WHEN TV MEETS THE "INTERACTIVE PROMISE"

Drawing on the definitions of interactivity, we might say that the "interactive promise" refers to the opportunity to participate in message production within a context of mutuality and relative equality. Interactivity thus depends on the accessibility of the message production process to all participants; it is the promise of "dis-alienating the consumer from the means of production" (Jarrett, 2008, p. 2). One might ask if this promise is always realized, even in a so-called interactive media environment. That issue notwithstanding, it is undeniable that technological developments have widened access to the means of media production. This access is closely related to cultural developments, such as those described by Jenkins's (2006) notion of "participatory culture."

While digital media can potentially enable broad access to the means of production, television remains mainly a mass, top-down medium that can not afford interactivity unless it transforms itself into a bidirectional or "many to many" medium. Traditionally, broadcast TV's transmission model is one-directional and centralistic (Einav, 2004), in contrast to interactivity, which "represents the ability not just to see and to hear but to be seen and heard" (Andrejevic, 2006, p. 393). Baudrillard (1985) goes even further, claiming that the design and structure of mass media prohibit a reaction from the audience, and no exchange of messages between producers and consumers of media is afforded. Therefore, the concept of interactivity traditionally related solely to digital online media, and was perceived as contradictory to the essential nature of TV. As Gilder (1994) put it: "Interactivity, almost by definition, is a computer function, not a television function. Making the boob tube into an interactive hive of theater, museum, classroom, banking system, shopping center, post office and communicator is contrary to the nature of the box."[2]

Nonetheless, even though television is a "one-way" medium, producers have craved interaction with the viewers, or at least the appearance of interaction, from the beginning. In the 1950s, the American children's program *Winky Dink and You* invited the viewers to "interact" with the characters by drawing on a transparent plastic sheet attached to the TV screen (Jensen and Toscan, 1999, p. 18). The following decades saw various efforts to develop technologies that afforded some kind of interactivity. For example, Warner developed the QUBE system in the 1970s, which allowed for feedback from viewers through a designated cable. But this and other technological experiments in the 1980s and 1990s gained only limited success, and were eventually taken off the market (ibid., p. 19).

Technological and cultural changes in the media environment, particularly the interactive affordances (Hutchby, 2001) of digital media, have renewed the debate concerning the perceived activity or interactivity of the television audience (Holmes, 2009). This time around, the question that has been raised is not whether viewers are active in the reception and interpretation of television texts in the spirit of Hall (1980) and reception studies, but rather how interactive the communication process offered to TV viewers really is. In view of the recent developments in the media, the subject of discussion is now the dialectics of power relations between TV makers and their audience. Relating to what she calls "a performance of power," Hill (2016) describes the ways in which viewers today can go against the grain of TV and adopt practices that defy how the TV industry measures its audiences by developing personal viewing modes through both legal and illegal technologies and services. Ross (2008) focuses on opportunities for participation, depicting television viewers' "tele-participation" through digital platforms as a part of the contemporary viewing experience, and Hartley (2011) dubs the TV audience DIY viewers who expect to take part in televisual texts through various digital platforms. Research into reality television refers to viewers' involvement in voting for their favorite contestants, or in proposing themselves as potential participants (see, for instance, Holmes, 2004, 2009; Murray and Oullette, 2008; Turner, 2010). In contrast, Mayer (2016) reminds us that the power still rests with the corporations, since only a few "prosumers" are actually allowed into the industry.

In a similar vein, the concept of para-interactivity takes into account the change of power relations between producers and consumers of media in the wider mediascape, along with acute awareness of the lack of a parallel fundamental change in the relations between TV and its viewers. Thus, the term *para-interactivity* relates to a cluster of strategies through which TV aims to imitate interactivity but does not actually fulfill the "interactive promise."

PARA-INTERACTIVITY

This book therefore introduces a new theoretical concept which I call "para-interactivity." As I will explain below, while the concept is inspired by TV's legacy of televisual strategies, it denotes a phenomenon unique to the current era.

Mapping developments and changes in televisual texts is an intricate and complex task, as the old meets the new in many different ways. In my efforts to analyze changes in contemporary television, I am greatly indebted to veteran researchers, Horton and Wohl (1956), who coined the term *para-social interaction*. These scholars deconstructed interpersonal, face-to-face interactions as implemented on television, which they dubbed "para-social relations," breaking them down into their component parts: eye contact, light conversation, body language, greetings, and reciprocal dialogue. They then traced these elements and their translation into televisual language, concluding that televisual para-social interactions simulate interpersonal interactions "yet incompletely," as Handelman (2003) noted, some crucial quality is obviously missing. Following the lead of Horton and Wohl (1956), I deconstruct the interactive communication process and trace the translation of its components into televisual strategies that simulate interactivity. But once again, in this instance some crucial quality is missing.

I propose the term *para-interactivity* as a concept that can facilitate a typology of contemporary as well as traditional strategies employed by television producers, and contextualize them in relation to a changing media environment. Para-interactive strategies are inspired by various different elements of the interactive communication processes nevertheless are closely interrelated.

The concept of para-interactivity encompasses three main televisual strategies in three dimensions.

Para-interactive paths

These paths invite viewers to interact with television by responding to, participating with, or making an impact on the text. They are intended to produce a seemingly interactive viewing experience, potentially enabling viewers to change the mediated content (Jensen and Toscan, 1999) or to participate in shaping the mediated environment (Williams et al., 1988). In effect, however, they create little more than the appearance of interactivity through structured and constricted paths of input which invite viewers to interaction in various positions in respect to the screen and the televisual text: **in front of the screen**, interacting with the TV set and additional devices and services in order to make choices and select the desired content and when and where to view it; **on the text**, providing input and impacting the text, narrative, and characters in a program; **in the text**, participating in a program; **alongside the text**, interacting with the program and accompanying texts through a second screen alongside the broadcast; **in front of the text or "intra-activity,"** examining the self while watching the program, such as comparing oneself to the participants or seeking self-improvement through the program.

Although these paths are not entirely new to TV, in the past decade they have become more prominent and varied as part of television's attempt to appeal to its viewers and promote engagement. Naturally, all programs on TV can be watched passively without any interaction whatsoever. Nevertheless, the para-interactive paths aim to paint the texts in interactive colors even for those viewers who choose not to take them. This strategy is discussed in length in chapter 3.

Temporal Para-interactive strategy

This strategy is meant to produce an ostensibly "live" or "real-time" viewing experience in order to create a temporal link between televisual texts and viewers. In addition, television produces "simulated live" programs, that is, taped shows infused with "liveness indicators." I have also identified two new modes of televisual "liveness": "second shift liveness" and "alternative liveness" produced through digital platforms that extend the appearance of liveness beyond the broadcast. The temporal strategy echoes "real time liveness" (Bennet, 2008), which affords real-time interactivity similar to that of a computerized system. It validates and paves the para-interactive paths by enhancing the "now-ness" of television texts, depicting them as still unfolding and therefore capable of being shaped and impacted by the viewers. This strategy is discussed in detail in chapter 4.

Spatial Para-interactive strategy

This strategy aims to unveil television's apparatus on screen in order to create the appearance of an inclusive viewing experience by participation in the production process and simulated reciprocity between producers and viewers. As television production has become more visible and transparent, the line between front and back stage has become increasingly blurred. Although this, too, is not entirely new to TV, it is now employed much more frequently.

Unveiling the production apparatus is achieved through the bidirectionality of discursive and visual elements: in one direction, cameras (and audiences) are invited backstage; and in the other, "backstage elements" are brought forward to become an integral part of the program. I contend that this strategy is inspired by the promise of a potential role exchange between participants in a process of mutual discourse (Williams et al., 1988), and by the assumptions of participatory culture, that is, a changed relationship between consumers and producers of media based on reciprocity and equality (Jenkins, 2006). However, it is still a **para**-interactive strategy, unable to produce genuine interactivity and role exchange between producers and audiences. Even though the production apparatus has become more visible on screen, most often it is only partially unmasked, revealing nothing more than a "staged backstage." This strategy is discussed in chapter 5.

I propose that as a conceptual tool, the notion of para-interactivity can help create a typology of both contemporary and traditional strategies employed by TV and contextualize them in respect to a changing media environment. Not only can it aid in explaining some of the recent changes in TV texts, but it also links the professional discourse of TV makers to the content they produce. Moreover, it is a cross-genre concept, so that the various para-interactive strategies can be found in a range of TV genres.

IDENTIFYING PARA-INTERACTIVITY: CORPUS AND METHODOLOGIES

The conceptualization of para-interactivity is based on research using a variety of complementary methodologies to analyze an extensive corpus. First, I conducted a qualitative analysis of prime-time (19:00–23:00) programs that aired on Israeli commercial television broadcast channels during the years 2008 to 2010, 2014, and 2016. Initially, all prime-time weekday programs were recorded during the period between 2008 and 2010 and watched closely at least twice. Due to the growing size of the corpus, in the years 2014 and 2016 only the prime-time programs aired on the first week of every month were recorded, watched, and analyzed. The analysis was aimed at identifying the various strategies employed by TV to appeal to its viewers and invite them to interact with the text or its producers. Based on the rationale that "television is no longer a stand-alone medium" (Turner and Tay, 2009, p. 7) supplementary video and textual content from digital platforms and related social media content was also analyzed. In addition, an analysis was conducted of around 60 journalistic articles published in the popular media, along with blogs and trade sites, which provided further understanding of the TV industry.

The second methodology entailed semistructured interviews with TV executives, producers, and lower-level industry employees in an effort to uncover some of the concerns and dilemmas of television professionals, as well as their perceptions of TV and its audience. This line of inquiry derived from the assumption that understanding the professional doxa (Thompson, 1990) of TV makers is significant because it impacts the televisual texts they produce (Lotz, 2004, p. 23). A total of 57 interviews were conducted in four waves: 2008–2009, 2011, 2013–2014, and 2016. The relatively extended duration of the research made it possible to explore whether there are changes over time in the industry's perceptions and discourse. The interviewees were recruited largely by means of the "snowball" method, with one interviewee recommending the next and sometimes helping me to contact them. Almost all interviews were recorded and transcribed for purposes of analysis. In the case of two interviewees who refused to be recorded, I took detailed written notes. Conducting interviews with people in the TV industry posed numerous obstacles, including lack of accessibility and disclosure agreements which led to their demand to remain anonymous. I am therefore unable to provide any data that might reveal the identity of the professionals cited in the book. The obstacles and opportunities I encountered in the course of the research are elaborated on at the last part of chapter 2.

Finally, I attended television industry trade gatherings in 2008, 2009, 2013, 2014, 2015, and 2017, where I conducted participatory observations. Trade conferences are considered "half-way spheres" between the public and the restricted production spheres (Caldwell, 2008), and therefore offer an opportunity for a closer look at the industry. Attending these six gatherings also provided extra benefits, since it enabled me to start informal conversations with TV makers and later recruit them as interviewees for my research.

Although my research focused on Israeli commercial TV as a case study, I strive in this book to offer a universal conceptual tool that can be applied to TV texts anywhere in the world where broadcast TV is facing similar challenges. That said, Turner and Tay (2009) note that TV is understood, used, and interacted with in a variety of ways around the world (pp. 3–5). Therefore, there is a need to investigate TV in other locations as well, potentially developing or adapting the tool I propose here. Another challenge this project confronted is the fact that TV, as a medium and as an industry, is undergoing constant changes sparked by the rapidly transforming media environment. As a researcher, I often had the feeling I was chasing a moving target. Even though there will most probably be additional changes after this book is published, I hope it manages to capture a moment in time in the ongoing transformation the media is experiencing, and will help in tracking the many innovations yet to come. This may require further development of the concept of para-interactivity as presented here. Last but not least, one element is still missing from this analysis—the viewers. My project investigated TV makers, their perceptions of the viewers, and the texts they produce. Future research might focus on understanding the audience's preferences, viewing practices, and reception of TV programs, an issue which is also of great interest and significance in these tumultuous times for the media.

NOTES

1. This data was presented by Jon Steinberg, founder and CEO of Cheddar TV, in INTV. Innovation TV conference 7.3.17, Jerusalem.

2. "Life After Television, Updated," http://www.seas.upenn.edu/~gaj1/tvgg.html (Accessed: 10/4/2009).

Chapter Two

"1.0 Producers in a 2.0 World"

The Dual Discourse of Television Makers

The title of this chapter is a paraphrase of a comment by author Zadie Smith in a review of the 2010 movie *The Social Network*. Written by Aaron Sorkin and directed by David Fincher, both in their forties at the time, the film tells the story of Facebook. Smith noted that "it's clear that this is a movie about 2.0 people made by 1.0 people."[1]

When I was interviewing television producers and executives for my research, I was often reminded of Smith's remark. All too frequently I had the feeling I was talking with 1.0 TV makers producing content in a 2.0 world, while trying to make sense of the changing media environment.

The Israeli television industry, like its counterparts around the world, is facing a range of challenges in a constantly changing mediascape. With viewer ratings dropping, broadcast channels are operating under the constant fear of losing their audience.[2] For academic researchers, on the other hand, times of change provide an excellent opportunity for a close examination of the changing roles and power relations in the television industry (Van De Ven and Poole, 1995).

Taking advantage of this opportunity, I set out first to examine the professional discourse in the industry between 2008 and 2016, critical years in terms of technological and cultural change. My research is based on interviews conducted in four waves (2008–2009, 2011, 2013–2014, 2016) with broadcast television producers and executives, along with participant observations at six industry gatherings,[3] and numerous relevant media articles and blog posts.

This chapter presents the results of that research, beginning with a short review of the filed; Israeli broadcast television and the challenges it faces.

This is followed by an outline of the dual professional discourse of TV makers regarding the role of television in a changing environment, their own professional skills, and the way they conceive their viewers and their changing expectations. Finally, I argue that this professional duality is the ground on which the various para-interactive strategies in contemporary television have developed. Last but not least, I share some afterthoughts and methodological insights gained in the course of my research, along with the obstacles and opportunities I encountered in the study of the television industry.

ISRAELI BROADCAST TELEVISION

The research presented in this chapter focused on the Israeli broadcast television industry, and specifically the people involved in making television: producers and executives (both senior and junior). As many readers may not be familiar with this arena, I shall begin with a short historical review.

Television came late to the Holy Land. The first TV station, a public channel, started broadcasting only in the late 1960s, funded by a mandatory license fee. For over twenty years, the channel enjoyed exclusivity, and played a significant social and cultural role in Israeli society. After the introduction of commercial television in the 1990s, however, the public channel lost most of its viewers. Organizational difficulties, intense political interference, and a lack of effective management led to negative public sentiment toward the IBA (Israeli Broadcasting Authority). In fact, in 2014, the Israeli government decided to abolish the license fee, dismantle the malfunctioning body, and establish a new public broadcasting agency.[4] Finally, following numerous postponements by the government,[5] of the state-run Israel Broadcasting Authority (IBA) was closed on May 9, 2017, and was replaced with a new, smaller entity known as Kan a few days later.[6]

The first commercially funded TV channel (Channel 2) was introduced in the early 1990s, and was joined almost a decade later by another commercial broadcaster (Channel 10). Israel adopted a hybrid model, creating a mix of PSB and a free market approach (Lachman-Meser, 1997). Hybrid broadcasters are, by definition, commercially funded on the one hand, and subject to laws defining their obligations and privileges on the other (Hellman and Keinonen, 2013, pp. 45–46).Thus, even though the two commercial channels in Israel are financed by advertising, they are still considered public assets and required by law to serve the public interest (Doron, 1998). Consequently, the private companies operating these channels for profit are expected to deliver certain social and cultural benefits to Israeli society, as defined in the Second Broadcast Authority Law (Sahaf, 2009), resulting in a situation in which cultural programming is embedded in commercially oriented channels (Katz, 2012).

Although the Israeli television industry is fairly small, it is competitive, vibrant, and innovative,[7] and has drawn international attention,[8] with original formats it has developed being exported globally.[9] Moreover, despite the fact that the commercial channels are constantly coping with financial hardships and frequent changes in personnel, their viewership is relatively steady. Israelis sit in front of their TV sets for an average of three and half hours a day, and are especially keen on watching the many news and current events programs aired every evening, as well as the numerous reality shows. At the same time, however, the Israeli public is deeply immersed in the digital environment, with Internet use reaching almost 80 percent.[10]

TV MAKERS' PERCEPTIONS

My interviews revealed that Israeli TV makers are very concerned about television's future as the dominant medium, and the industry is facing considerable challenges in the changing environment. A challenging time for industry workers is an interesting time for researchers. In a period of cultural and industrial change, power relations erode and new opportunities to reexamine the norms of the organizational structure present themselves (Van De Ven and Poole, 1995). In respect to the production of culture and media, Thompson (1990) highlights the significance of regarding the ways in which content is produced as inseparable from the content itself. He therefore recommends that media researchers analyze the doxa, hence the perceptions, opinions, and understanding of the people who produce the messages in order to understand the content they produce (p. 279). In this spirit, I undertook to explore and analyze TV makers' perceptions of three main issues: television's role and relevance in the changing media environment; how TV professionals perceive themselves and their skills in a changing industry; and their perceptions of the viewers' preferences.

On the whole, the professional discourse can be described as dual and ambivalent. On the one hand, it expresses confidence in television's unique cultural role and, as a result, it's continuing relevance and popularity, while on the other hand it voices deep concerns about television's slow adaption to the digital world and the fear it will lose its audience. The same duality also characterizes TV makers' discourse on their own "tools of the trade" in a changing environment and their prospective careers in the industry. This pronounced duality is even more intriguing given that the two conflicting views were often expressed by the same person in a single interview.

The interviews centered around the following issues:

1. Should TV change in the digital environment? Is TV still appealing as a medium in the digital age? Should TV adapt to the changing environment? What is the future of TV?
2. Should the professional skills of TV makers change? Do younger employees have a better understanding of the digital environment?
3. What do viewers want? What are the preferences of audiences in the digital age? How can TV appeal to the viewers?

SHOULD TV CHANGE IN THE DIGITAL ENVIRONMENT?

In the last few years, numerous academic books and panel discussions at academic conferences have been titled something along the lines of "Is Television Dead?," "Is Television Nearing Its End?," or "The End of Television?" (Katz and Scannel, 2009). At industry gatherings, TV professionals attend debates on the grim fate of television. At an annual industry gathering in November, 2008 (Rosh Pina Television Festival), for example, one panel discussion was titled "The Battle in the Living Room." The moderator opened with the words: "Every year I chair a panel predicting the death of television."[11] His remark captures the ambivalence of TV makers, alluding to their fear of losing their audience while at the same time ironically expressing criticism of the time-worn predictions of the death of the medium.

Although few years have passed since that conference, even in the interviews I conducted in 2016, the professional discourse has remained dual. TV producers and executives typically express confidence in television's advantages over digital media and, as a result, the continued appeal to its viewers, while at the same time voicing deep concerns about its slow adaption to the digital world and the viewers' desire to participate in the production of media in a manner that traditional TV can not offer.

TELEVISION'S ADVANTAGES AND ONGOING RELEVANCE: "I AM ONE OF THOSE WHO BELIEVE THAT TELEVISION WILL PREVAIL."

Although they are aware of the challenges that TV is facing, people in the industry emphasize the strengths of traditional linear broadcast television, in particular passive viewing, TV as storyteller, and the opportunity for shared viewing.

Passive Viewing

The familiar distinction between TV as a "lean back" medium and the computer or Internet as "lean forward" (Uricchio, 2004) is invoked in the professional discourse in depictions of television as a leisure activity, as opposed to the Internet which is "work related."

> In the final analysis, viewers want to lean back and be passive . . . they don't want to be active all the time. Television is leisure not work. (Interview with M., Jan. 27, 2009)

> I use the Internet for work. It does not replace TV. (Interview with D., Dec. 7, 2008)

Moreover, TV makers' perceive of the medium as a much needed "editor" of content.

> What is missing online is editing. Someone who will sort things out because I don't even know where to begin, I don't want to work, I want to have fun. (Interview with P., June 6, 2009)

Indeed, one of my interviews took place in the office of an executive at a commercial channel (Dec. 7, 2008). As he was explaining the advantages of TV, he suddenly turned to his computer screen and Googled something. He then turned back to me triumphantly and said: "You see?" I was actually sitting too far from the computer to clearly see what was on the screen, but I could tell he was pointing to the page of results his search had produced. "You see? This is exactly what people don't have the patience for . . . millions of possible results, content, data, and you have to find the best ones yourself. When you watch TV there are people who do that for you." Obviously, he was right. TV does much of the work for its viewers: selecting and scheduling programs and inviting them to lose themselves in an endless flow of images and stories. As my interviewee, an executive in his late forties, gazed at the screen with a weary expression on his face, he seemed like a typical "digital immigrant" (Prensky, 2001). He could find his way in the digital world, but it was a demanding task for him that required effort. He felt more comfortable in the televisual environment where "there are people who do that for you." His words appeared to echo Ellis's much earlier observation regarding broadcast television's liberating effect for viewers who struggle with "choice fatigue and time famine" in the face of abundant media texts and images, and its ability to alleviate the constant stress of feeling you are missing out on something (Ellis, 2002, p. 176).

TV as Storyteller

Another advantage of broadcast television cited by industry professionals is that it serves first and foremost as a storyteller. Fiske and Hartley (1978) portray TV as the bard of our time, claiming that it fulfills the bardic function by constructing the stories that comprise our culture and the reality within which we operate (p. 85). Quite a few of the TV makers I interviewed highlighted this feature of television.

> What viewers are always looking for is a good story. . . . They want someone to tell them stories. They don't want to tell the story by themselves. (Interview with M., Jan. 27, 2009)

Another interviewee took this claim even further.

> People watch TV because they're always looking for stories, information, must-see events. . . . People want to sit and watch someone else tell the story, and television can do it best. (Interview with D., Dec. 7, 2008)

This is an indirect reference to digital interactive platforms or social media, where audiences can take part in creating the media message and tell their own stories. It represents the view that despite the allure of the web and social media, television can still "captivate" its viewers by telling them a good story (Sconce, 2004, p. 95). It also alludes to the viewers' preference for passive consumption of media content and "stories," as opposed to the effort needed to post online. Storytelling was also mentioned as an important universal skill for TV makers, as we will see later in this chapter.

Shared Viewing

A further unique strength of television, according to people in the industry, is its ability to construct shared viewing, whether on the level of family, community, or nation. As one producer commented:

> I still see people's strong desire to sit and watch the same program with other people . . . not only here in Israel but in other places in the world too. (Interview with K., Sept. 7, 2009)

Television has always been considered a medium that can afford people the sense that they are together even when they are separated geographically (e.g., Dayan and Katz, 1994). Ellis goes even further, describing broadcast television with its fixed schedule as a uniting force, in contrast to digital media which fosters individuality. Broadcast television provides a voluntary point of social cohesion, of being together while being apart. Most of the models of interactivity and choices assume a lone consumer making choices in isolation. The limited arena of TV scheduling provides the basis on which households can negotiate more collective choices (Ellis, 2002, p. 176). The viewing experience can be shared spatially, such as when a family watches television together in the living room, deciding collectively which program or channel to watch, or temporally, when a group of people or a whole nation watch the same thing at the same time even though they are not in the same place. Linear real time television encourages shared viewing, and TV makers stress the importance of this feature.

> Broadcasters want to invite the whole family. . . . The programs enable the whole family to watch. . . . It's a shared experience. Unlike online content, television is still something that you talk about with others. (Interview with M., Jan. 27, 2009)

The industry people I interviewed consider the Israeli public to be more keen on shared viewing than people in other parts in the world.

> Here in Israel people want to feel they belong. . . . We can be at war anytime, so we feel the need to know what's going on, to belong, a need to huddle together . . . [and] watch TV together. (Interview with S., Dec. 14, 2009)

While Israel is definitely unique in terms of the geopolitical circumstances which might encourage shared television viewing, especially of news and current events, the fact remains that elsewhere in the world, broadcast channels still attract viewers to linear TV as well (e.g., Scandinavia: Kjus 2009a; 2009b; 2009c).

So far so good. As we have seen, the professional discourse presents television as a strong and irreplaceable medium that affords a much-wanted passive viewing experience by fulfilling the functions of editing and storytelling and enables the shared viewing that people need. Nevertheless, almost all interviewees admitted that given the changes in the media environment, television also has to change.

TV HAS TO CHANGE

> "Television will not die but it needs people who can look technology in the eye and adapt to it." (Interview with R., March 9, 2016)

Threatened by instability in the industry, TV professionals try to rationalize the state of uncertainty (Caldwell, 2004b, p. 45). They articulate TV's strengths as a medium and relevance as a cultural institution, while at the same time urging the industry to adapt to technological and cultural changes.

> The biggest fear hovering over this place is its commercial model. Naturally, we are looking for new models and they will be found eventually. I think we're now a moment too early . . . to understand how to make money from the expensive content we produce. (Interview with K., Feb. 27, 2016)

> In every meeting I say let's see how we can bring the internet to us. How do we bring the audience to us? (Interview with D., Dec. 25, 2008)

> The interactive thing is especially important. It can give the viewers, surfers, and listeners more options to express themselves. (Interview with Y., Jan. 11, 2014)

We can see that in the minds of TV makers, adapting to the digital environment means bringing the Internet to TV in order to bring the audiences with it. In other words, it is the "interactive thing" they need. However, precisely how to do this is unclear to most of them, and at least some are willing to admit TV's limitations and TV professionals' shortcomings in this regard.

> We use online platforms. But it's still TV coming to the Internet and not vice versa. . . . For a long time we have been stuck in the same place and that's because TV is a slow and cumbersome medium. . . . The technology, the people . . . we're stuck on the things we know. (Ibid.)

It would be reasonable to assume that professional discourse relating to the ways in which TV needs to adapt would become more focused and structured and less frustrated in the more recent interviews, as TV makers themselves adapted and acquired the relevant skills. However, this was not the case. The perception that television is not adapting to the media environment fast enough was even more pronounced in the most recent interviews.

> Television will not die, but it needs people who can look technology in the eye and adapt to it. I think that most of the TV executives today don't know how to do it. They're used to their old tricks and attitudes. Until one of them goes out into the new world, experiences technology, and then comes back [to television] and says we have to change, nothing will change. . . . Old habits are hard to change. (Interview with R., March 9, 2016)

Thus, TV makers acknowledge that one of the main reasons for the slow adaptation of television to the new environment is their own difficulty adopting the new technologies and abandoning their old habits. Moreover, since "TV workers shape and refashion their identities in the process of making their careers in industries undergoing political transitions and economic reorganizations" (Mayer, Banks, and Caldwell, 2009, p. 2), exploring the ways they perceive their professional skills is highly relevant to this discussion.

TV MAKERS' SKILLS IN THE DIGITAL AGE

The duality that characterizes industry discourse on the future of television is also present in the discourse relating to the "tools of the trade" in a changing mediascape. On the one hand, some professionals claimed that making TV today requires the same skills as ever, while others regretted their lack of "digital tools." The interviews revealed a confluence of two media cultures that sometimes took the form of a clash, often between different generations.

"You will always need the same skills"

Many TV makers insisted that industry professionals today need the same skills as ever.

> There are some skills you need no matter which platform you work with. These are interpersonal skills, intuition, curiosity. (Interview with L., Feb. 6, 2016)

> The main thing is to know how to tell a story—all the rest are techniques. If you can tell a story or communicate information you can do television anytime. (Interview with D., Dec. 7, 2008)

Some professionals claimed the cultural capital they draw on goes even further back, to cinema. I interviewed P., for example, the morning after the first episode of the second season of *Survivor* on Israeli TV (Dec. 24, 2008). An executive at a commercial broadcast channel in her early forties, P. was in high spirits and her office looked festive with colorful balloons clinging to the ceiling. She was keen to talk about the program that had aired the previous night, and explained that she was given the balloons to celebrate the high ratings it had received. I had seen it, and was impressed with the elaborate opening scene of helicopters carrying the contestants to the remote location on an exotic island. I complemented P. on it, and remarked that it was probably expensive and complicated to produce. She agreed, saying: "Yes, but it was worth it." I assumed she was referring to the ratings, but to my surprise she went on to explain that the whole scene was her homage to the film director Francis Ford Coppola and his famous helicopter scene in *Apocalypse Now* (1979), adding that it had been her dream to do something like that ever since she studied cinema at university. As she described the similarities between the original 1979 cinematic scene and the impressive opening scene of the reality show, it was evident she was proud that her skills drew on her knowledge of classic filmmaking. Thus, even in a unscripted show on a commercial channel, she aspired to integrate cinematic references and style. Apparently, at least for some people in the television industry, mastering cinematic style and aesthetics is an indication of cultural capital in the spirit of Bourdieu (1996), enhancing their status in the professional hierarchy. This attitude might derive from the fact that Israeli commercial TV was founded by graduates of the Tel Aviv University School of Film (now School of Film and Television), and many of them still hold influential positions, thus shaping production norms and tradition for the whole industry.[12]

P.'s comments represent one side of the ambivalent discourse relating to the skills of TV makers. On the other side are claims that, like the medium itself, industry professionals need new skills to adapt to the way of thinking of digital media.

The Generation Gap

The changes in technology and the media environment appear to have created a generation gap in the TV industry, much like that in society at large. Among other things, this state of affairs impacts TV maker's notions about their own skills.

> We are in transition now. It is a long transition period. When the teenagers of today grow up and go into the industry, it will be simpler. They have the skills, they can identify the needs, they live in the digital world, not like me. (Interview with D., Dec., 7, 2008)

While the older generation of TV professionals value their broad cinematic and televisual knowledge, they are also aware of their limitations.

> Television has to change, but in the meanwhile most of the people in TV "think television," that's their problem. If we want to change we have to think new media. (Interview with D., Dec., 25, 2008)

It is not entirely clear here whether D. believed she had to change or whether she was referring to her colleagues. Whatever the case, her comments raise another question: What exactly does it mean "to think new media"? As the digital department manager at a commercial channel explained, many TV professionals do not have a clear understanding of what is needed.

> They don't know exactly what they're asking. They say they want us to produce something digital, something viral. . . . They don't know what. They want us to "just do some digital magic somehow" without knowing what the possibilities are. (Interview with N., Jan. 23, 2014)

All these remarks reveal the concerns and frustrations of TV workers regarding their "tools of the trade." Like members of other industries going through a process of change, they show a tendency to try to maintain their familiar and well-established identity and methods of operation (Weick and Quinn, 1999). At the same time, however, they are acutely aware that younger employees bring with them a much needed "new media thinking," which some veteran TV executives have trouble understanding. In the words of one executive in his late fifties:

> When I hire new employees they have to be cross-media, trans-media, multi-system, havc a wider range of thinking. . . . They have to think about all the other new media. (Interview with Y., Jan. 11, 2014)

Indeed, Israeli broadcasters today have what they call "digital" or "interactive" departments that develop and manage the various online platforms and social media. According to one of my interviewees, the employees in this department in his company were referred to (outrageously) by a female executive as "the young digital girls in shorts" (interview with N., Jan. 23, 2014).

The ambivalence toward younger employees who represent the "new media culture" in the television industry reflects a deeper cultural and organizational tension. This is illustrated by the following anecdote.

At the end of my interview with M. (Dec. 14, 2009), an executive in her early forties at one of the commercial channels, she escorted me to the elevator along a hectic corridor. We were still talking when I noticed the head of the interactive department walking past. I asked M. if they shared the same floor with her department, and she said they did. A moment later she turned to me and whispered with a half smile: "One day they'll probably take over the entire floor, if not the whole channel." Her unexpected comment gave me a glimpse into the organizational and professional tensions in the broadcasting company. It was particularly surprising as it came after an hour-long interview in which M. insisted that television will prevail and that digital platforms pose no threat. The incident shed light on relationships in the organization, which seem to teeter on the edge between cooperation and competition. Moreover, it highlights the pressure on TV makers' professional identity and their insecurity and ambivalence toward the major changes their workplace is undergoing.

The duality that characterizes professional discourse regarding TV's future and TV makers' professional skills does not extend to their perceptions of audience preferences. In this respect, there appears to be a general consensus that the viewers have changed in the wake of the new media environment.

"VIEWERS TODAY ARE DIFFERENT"

Media producers routinely try to imagine their audience in order to produce content that will appeal to them. In times of change, this task can become quite confusing. Nevertheless, my research revealed that most TV professionals agree that the viewers are different and the whole experience of media consumption has changed.

> Viewing is already different and will change even more. Kids today consume content on their computers and other devices. . . . They don't even have the patience to watch a full-length movie. When they're watching a program, they're doing other things at the same time, surfing, texting. . . . I call this the twenty-first century attention span. (Interview with P., Dec. 25, 2008)

When I asked exactly what viewers today want, people in the TV industry said they want to participate and to feel they are part of the televisual text and its production. Moreover, they often linked this desire to wider cultural and social change, and the fact that people no longer merely consume media content.

> Communication between people has changed. You don't want to just sit around and laugh together. Nowadays you need Twitter to tell [everyone] what you're doing, you need a Facebook community to feel that you're seen, noticed. (Interview with K., Sept. 7, 2009)

> We're in the age of narcissism. Everyone wants to be heard somehow. . . . Lots of people have something to say . . . and television is part of it. TV is still a mass medium, but we're listening to the changes in society. (Interview with S., Dec. 14, 2009)

Despite any reservations they may have regarding the new media, TV makers appear to be quite adept at identifying the changes in the media environment and their impact on TV viewers' expectations.

> The popular programs, those with high ratings, are the ones in which the viewers feel they're part of something, part of the celebration, part of the production, part of behind the scenes . . . maybe part of the televisual creation. (Interview with K., Sept. 7, 2009)

> Young viewers especially want to be part of everything, an active part, unlike our generation or our parents who were willing to be passively "spoon-fed" by TV. (Interview with D., Dec. 25, 2008)

> There is a correspondence between the technological possibilities and the viewers' expectations. . . . They want to be more active and they actually have more options. The audience not only decides what to watch but also how the storyline will develop. (Interview with Y., Jan. 11, 2014)

Thus, although television producers and executives may not agree as to whether or how TV has to change, and their professional skills along with it, there is widespread agreement among them that the viewers have changed and the ways they watch TV is different. The TV makers I interviewed claimed that viewers want to feel as if they are active and interactive, that they can impact and participate in the programs and be part of the production process.

CONCLUSION: WHAT DOES IT ALL HAVE TO DO WITH PARA-INTERACTIVITY?

The TV industry today is coping with changes in technology, culture, and expectations. As we have seen, TV makers' discourse is marked by duality and ambivalence toward the medium's relevance and need to change, as well as the necessary professional skills and cultural capital. Television is therefore an industry in transition, where professional and aesthetic cultures meet and often collide, thus producing organizational tensions.

Furthermore, in this time of profound change, television professionals believe viewers' expectations are also changing. They express their conviction that viewers today seek involvement in the production process, as they are used to an interactive communication environment in which they can easily become producers and distributors of media content, often from their own homes. Indeed, the very definition of interactivity relates in part to the degree to which participants in a communication process can exchange roles in their mutual discourse (Williams et al., 1988). When it comes to television, however, the medium's different affordances (Hutchby, 2001) make this highly unlikely. Television is still mostly a linear medium that does not offer viewers the opportunity to trade places with TV makers and become producers of TV content (Ellis, 2004, 276).

In times of change, aesthetic choices and creative style are reexamined and reformulated. Caldwell identified an encounter between two competing perceptions in the American TV industry in the 1980s: cinematic culture and aesthetics versus the perception of TV as an antithesis to film. This duality eventually produced a new style which Caldwell conceptualized as televisuality (1995). In the same spirit, I identify two competing media cultures in today's TV industry: television culture versus digital culture. In order to adapt to the interactive new media environment and to what they perceive to be the changing expectations of their viewers, TV makers are now producing a synthesis of television content which integrates elements deemed to belong to the interactive digital culture. Their colliding and intersecting perceptions have given rise to a variety of narrative, discursive, and visual strategies which I conceptualize as para-interactive. That is to say, these strategies imitate interactivity within the limitations of the linear medium of TV. Para-interactive strategies are aimed at creating a viewing experience that invites the audience to feel as if it is part of the text and its production.

The specific para-interactive strategies are discussed in depth in the following chapter. However, before moving on, I would like to take a closer look at the research process that produced the findings presented in this chapter and share some afterthoughts. Although my study did not fully meet the definition of production research, it nevertheless posed similar challenges.

AFTERTHOUGHTS: OBSTACLES AND STRATEGIES IN TELEVISION PRODUCTION RESEARCH

Production research is a relatively young member of the television studies family. It aims at collecting empirical data on the complexity of production routines, as well as on professional, economic, and cultural forces that influence the production process and shape television content.

We have recently been witnessing a renewed interest in television production, and quite a few books have been published on the subject (Caldwell, 2008; Havens and Lotz, 2012; Lotz, 2007; Mayer, Banks and Caldwell, 2009, among others). One possible explanation for the reawakening of production research is the rapid transformation that television is undergoing. The current changes impact not only production, but also the industry's relationships with advertisers and viewers alike. As an industry in flux, television is a fascinating field for production research, which can provide invaluable insight into an industry "on the move."

My research plan called for interviews with professionals in Israeli broadcast television and observations of the production process in order to learn about the industry's doxa and praxis (Thompson, 1990). It did not take long for me to realize that carrying out this plan was going to be more challenging than I expected. Eventually, I managed to interview nearly sixty television executives and production workers to perform several observations of the production process and industry gatherings, all which produced invaluable data.

Below I present the major obstacles I encountered in the course of my research, along with strategies I developed for overcoming them. It is my hope that this prior knowledge of what they can expect will assist future scholars and encourage them to venture into the world of television production research.

Obstacles and Challenges

Accessibility

It is quite difficult to gain access to television executives and professionals. This is a well-documented challenge (Caldwell, 2008; Ortner, 2010; Shahaf,

2009), and I encountered it repeatedly during my research, for example, when the people I contacted did not answer my emails and calls, refused to meet with me, or agreed to an interview and cancelled at the last minute. There are many reasons why TV industry workers are reluctant to be interviewed. As a rule, they are extremely busy and their schedule is erratic. Moreover, some have celebrity status and unfortunately, many do not appreciate academic research and consider it a waste of precious time.

Concealment and Confidentiality

Almost all commercial television employees in Israel are required to sign a confidentiality agreement as a condition of employment (Klein Shagrir, 2015). One obvious reason is that the television industry operates in an environment of fierce commercial competition. In addition, international formats acquired by Israeli broadcast channels generally demand confidentiality. Furthermore, concealment plays a role in many unscripted television programs. In fact, a great deal of effort is invested in these so-called reality shows the television industry not only manufactures illusions but invests great efforts in producing these illusions of reality. One of the intriguing products of contriving illusions of reality is the "transparency/confidentiality paradox," an issue I relate to in further detail below.

Public Relations Smokescreens and Lack of Reflexivity

During interviews, television executives tend to cite PR messages that promote themselves and their companies. This makes it hard for researchers to distinguish between the façade and the professionals' genuine ideas and perceptions. PR smokescreens make it even harder to identify the challenges facing the television industry, since they typically express the utmost confidence in their company's continued success and mask any difficulties it is contending with. What is more, many television people in general, and senior executives in particular, hide behind public relations smokescreens because they are reluctant, or even unable, to discuss their work reflectively. I often heard an interviewee state that a certain process was performed because "that's the way it's always done" or that they went by "a gut feeling." This is understandable since in interviews people are asked to reflect on things they may take for granted (Hesmondhalgh and Baker, 2011). Moreover, reflecting on one's working process may be even more stressful in times of change and uncertainty. Nevertheless, such replies are frustrating for a researcher.

Love-Hate Relations with Academia

Based on my experience, some television professionals have a complex relationship with academia and academics. During the interviews I conducted, several of the interviewees declared openly they do not "see the point in academic research," and complained about it being slow to publish and detached from the "real world." Others tried to demonstrate that they were familiar with academic concepts and theories. In both cases, reacting to such comments can be counterproductive for the researcher, as it can divert the interview in nonrelevant directions. Surprisingly, or perhaps not given the current turmoil in the industry, even after voicing harsh criticism of academia, several interviewees inquired as to their chances of getting a university teaching job.

An Era of Change and Uncertainty

Rapid technological, cultural, and economic changes make it harder for television executives and professionals to draw a clear picture of the TV industry. The turbulent period it is going through directly impacts the workers and produces a state of precarious employment (Gill, 2010). In fact, one of the interviewees had worked for four different employers over the course of two and a half years. I realized that in times of uncertainty, it can be hard for media employees to communicate a clear representation of their professional perceptions and beliefs, as well as their work routines. Consequently, acquiring an accurate picture of the television industry can be challenging for the researcher.

Strategies: the Seven S's

As I sought to contend with the obstacles and challenges, I discovered several strategies that might prove helpful for other television production researchers as well. I call them the "Seven S's."

1. "***Study Sideways***" (Ortner, 2010)
   Look for intellectuals in the industry and professionals in academia. In other words, in order to gain access to industry professionals, try to find television professionals among your colleagues at the university, and look for people who have an academic background and now work in the industry. These individuals are not only more accessible, but can also be expected to be more reflective. In addition, I discovered that interviewing people outside the broadcasting companies who work with the television industry, such as regulation officers, can be very helpful. Such officials are more accessible than TV producers and have valuable insight into the industry.
2. ***Study Down***
   Opt for low-level employees. They are more accessible, less prone to quote PR catchphrases, and as a rule they are the ones who really know what is happening down on the production floor.

3. *"Snowball" Interviews*

   Especially in a small country like Israel, where the television industry is relatively limited in size, the snowball technique can be extremely useful. Once you get your foot in the door, people will help you talk to their colleagues. I must note, however, that in some cases interviewees were reluctant to refer me to their colleagues out of fear of breaking their confidentiality agreement.
4. *Select the Interview Location Carefully*

   The location of an interview is extremely important. Some high-ranking executives prefer to meet in their office, which has its advantages: less chance of cancellation and more flexibility with regard to the duration of the interview. In addition, holding the interview in an executive's office can sometime provide an opportunity for a glimpse at the atmosphere and interrelations in the organization. However, office interviews also have their shortcomings: they can repeatedly be interrupted by coworkers and are more likely to invoke official PR catchphrases.

   On the other hand, conducting an interview in a neutral place, such as a coffee shop, can produce an informal atmosphere and facilitate the interaction. Outside the office, however, some interviewees appeared to be paranoid about the people sitting nearby. One even introduced me as his cousin to a colleague who approached our table.
5. *Squirm Your Way into Industry Conventions and Gatherings*

   These events are "half-way spheres" (Caldwell, 2008) where participants speak more freely than usual.
6. *Sign up for Live Broadcasts and Tapings of Television Programs.*

   As an invited member of the audience, you can join behind-the-scenes tours that offer a chance to observe (albeit in a strictly limited manner) the television production apparatus.
7. *Surf TV Industry Websites and Subscribe to Television Blogs*

   These sites are surprisingly rich in information, and some bloggers even appear to have access to inside information.

The Transparency/Confidentiality Paradox

As I discuss at length in chapter 5, many programs aired on television today are accompanied by "behind-the-scenes" programs, promos that include backstage footage, or additional multiplatform content and news coverage that takes the viewers behind the scenes. Prime time programs also often integrate traditional backstage elements into the broadcast itself, such as auditions or editorial meetings. It would appear that the TV apparatus has become more transparent than ever before (Klein Shagrir, 2015).

Paradoxically, however, as television's backstage becomes more visible on screen, the industry itself is increasingly being concealed behind real and legal walls of confidentiality. Not only are television performers and employees compelled to sign confidentiality agreements, but even ordinary people who audition for a reality show, as well as backstage visitors, are often asked to sign an agreement before being allowed to enter the production area.

Against the backdrop of recent cultural and technological changes, the line between producers and users/viewers, between the television world and the ordinary world, and between professional TV actors and "ordinary people" is becoming blurred (Turner, 2010). Yet even though television's rhetoric invites viewer participation and celebrates the accessibility and transparency of the production, the TV industry is closing itself off to external view. Producers create a simulated "backstage" visibility on stage, while maintaining strict concealment and secrecy behind the scenes. This paradox can be misleading and frustrating to production researchers who do not recognize it for what it is.

Why is Television Production Research Important?

Media production, and more specifically television production, is just as significant as any other cultural production process. Moreover, as Mayer, Banks, and Caldwell assert: "We frequently come to know about media producers and their work, ironically, through the representations they make" (2009, p. 1).

The television production process impacts the design of central cultural texts, namely TV programs. Studying the economic and organizational forces and interests in the television industry is crucial to understanding these texts. Moreover, television texts reach wide audiences, and the celebrity status of people on and in TV, as well as the major economic and political interests involved in the industry, make it not only a fascinating field for research, but also socially significant.

In addition, as academics, we are responsible for the education of future television professionals. Understanding the production process will enable us to equip our students with tools for critical thinking that can be applied both to media texts and to the production process itself.

And finally, the television industry is transforming before our eyes, and TV makers and academics alike are striving to predict its future. As media scholars, I believe it is incumbent upon us to closely observe the changes and developments as they unfold.

NOTES

1. Zadie Smith, "Generation Why, "*The New York Review of Books*, Oct. 25, 2010. http://www.nybooks.com/articles/archives/2010/nov/25/generation-why/?page=1 (accessed: Oct. 10, 2011).
2. http://www.ynet.co.il/articles/0,7340,L-4528895,00.html (accessed: Sept. 9, 2016).
3. Rosh Pina Television Festival: 2008, 2009; INTV (Keshet) 2012, 2013, 2015, 2017.
4. http://www.the7eye.org.il/100086 (accessed: April, 4, 2017).
5. http://www.jpost.com/Opinion/Grapevine-IBA-closure-postponed-yet-again-460849 (accessed: July 20, 2016).
6. http://www.bbc.com/news/world-middle-east-39869757 (accessed: May, 22, 2017)
7. http://www.haaretz.com/print-edition/features/can-israeli-tv-shows-go-the-way-of-high-tech-1.367987 (accessed: April, 4, 2017).
8. http://forward.com/news/israel/211977/israeli-television-hitmakers-take-hollywood-by-sto/ (accessed: September, 20, 2016).
9. http://www.haaretz.com/business/israeli-tv-shows-are-already-a-hit-abroad-but-producers-believe-the-sky-s-the-limit.premium-1.514654 (accessed: April, 4, 2017).
10. All data from *Annual Report: The Israeli Media in 2015—Agendas, Uses and Trends*, Mann and Lev-On.
11. Danny Peled of Goome Interactive, Nov. 17, 2008
12. See Yotam Feldman, "Ari Fulman's Winning Cycle," *Ha'aretz Magazine*. http://www.mouse.co.il/CM.articles_item,1124,209,33253,.aspx (accessed: March 17, 2011).

Chapter Three

Para-Interactive Paths

Inviting the Viewers to Interaction

At the center of "The Garden of Forking Paths," a short story by the Argentinean author Jorge Luis Borges, is a fictional novel through which one can proceed by a variety of paths. The reader is invited to choose one or more paths, with each choice producing a different storyline that unfolds in a different timeline. Furthermore, the reader can choose more than one path at the same time and take several narrative paths in parallel. It has been claimed that "the concept Borges described in 'The Garden of Forking Paths'—in several layers of the story . . . is that of a novel that can be read in multiple ways, a hypertext novel" (Wardrip-Fruin and Montfort, 2003, p. 2).

The ability of the reader to choose the paths they take and thereby create their own story and reading experience is thus similar to the use of hypertext in a digital environment. Users can surf through links and sites, interacting with texts and audiovisual content and taking a different path each time. Applying this idea to the realm of television, I propose that many televisual texts today, inspired by digital media, offer viewers the opportunity to choose one or more paths through which they can presumably interact with the text or impact the narrative. However, whereas in a truly interactive environment the users' choices can have real impact, TV cannot provide genuine interactivity. As a linear and mostly one-way mass medium that is operated by large professional organizations and whose production process is complicated and expensive, television's affordances (Hutchby 2001) allow for only para-interactivity, or an imitation of interactivity. This para-interactivity is meant to create the impression (rather than the reality) that the viewers can be part of the televisual text, participating in it, impacting it, and engaging with it.

As explained in chapter 1, para-interactive strategies are inspired by the current perception of media audiences as producers, and not merely consumers, of media messages. They reflect TV makers' belief that nowadays television has to appeal to its viewers by inviting them to take part in the text and production. In this chapter I describe the various para-interactive paths in use today and the strategies they employ.

WHAT ARE PARA-INTERACTIVE PATHS?

Para-interactive paths can be defined as a cluster of televisual strategies that invite viewers to partake in potential or actual interaction. Structured into televisual texts and accompanying para-texts, such as promos, ads, and online content, they are meant to imitate hypertext, making the televisual text appear to afford the viewers the opportunity to shape and enhance their viewing experience by interacting with the text. A range of options is offered, and viewers can select one or more. The idea of range of options is closely tied to interactive communications, and is, in fact, part of some definitions of interactivity (Durlak, 1987; Kiousis, 2002; Steuer, 1992). According to Wu (1999), for example, "[Interactivity] can be defined as a two-component construct consisting of navigation and responsiveness" (p. 6). Thus, in order for interactivity to exist, there must be several options for interaction and activation so the user can navigate between them. Moreover, interactivity entails a certain degree of responsiveness to the users' choices and inputs. Many of texts on broadcast TV are therefore constructed to appear as if the viewing experience can be interactive in both senses: with a range of options for viewers' input and interaction as well as a guise of responsiveness to this input.

As a linear mass medium, television has not traditionally afforded actual interactivity, that is, it has not responded wholly and immediately to viewers' choices, excluding the changing of channels. Para-interactivity does not substantially modify this situation, offering only the appearance of interactivity through structured and constricted paths of input. It is only very rarely that viewers can actually impact a TV program. Moreover, some of the para-interactive paths are not entirely new to television. Recently, however, they have become more prominent and more varied as part of TV's attempt to appeal to viewers in the changing media environment of the digital age.

It is worth noting that, as always, TV programs can be watched passively, without any interaction whatsoever. Nonetheless, para-interactive paths aim to paint televisual texts in interactive colors, even for those viewers who choose not to take them.

PARA-INTERACTIVE PATHS

Para-interactivity can take various forms. On the whole, they can be divided into the following paths:

1. ***In front of the screen***: inviting viewers to interact with the TV set and additional screens and services in order make choices and select the desired content and when and where to view it.
2. ***On the text:*** inviting viewers' input and impact on the text, including the narrative and characters.
3. ***In the text:*** inviting viewers into the televisual text, that is, to become participants in the program.
4. ***Alongside the text:*** inviting viewers to interact by means of a second screen where they can consume more information, carry on discussions with other viewers, and participate in activities related to the text.
5. ***In front of the text, or "intra-activity":*** inviting viewers to engage in internal activity, such as self-examination or self-improvement, by comparing themselves to participants on screen.

FIRST PATH: "IN FRONT OF THE SCREEN"

This path invites viewers to determine what to watch, when and where to watch it. Thanks to technological developments, viewers today have a much wider range of options for choosing the content they want and the way to consume it. This type of interaction between people and machines might better be termed reactive than interactive, since it involves the machine's response to a simple input, such as the press of a button. Nevertheless, it enables the individual a degree of control over the media environment.

The first significant technological development that allowed consumers greater control over their viewing was the remote control. Invented by Zenith in the 1950s and dubbed the "Lazy Bones," it was originally connected to the TV set by a cable. The introduction of the remote control enabled viewers not only to change channels and select the program they want to watch more conveniently, but also gave them the ability to skip commercial breaks by switching easily back and forth between stations. With the subsequent invention of home video recorders, media consumers were able to control not only what to watch, but when to watch it, and even allowed them to skip the commercials altogether.

In the 1990s, the world waited with baited breath for the introduction of interactive TV (ITV). Hailed as "two-way" television in which viewers can select content and their input can be sent to broadcasters (Jensen and Toscan,

1999), ITV was expected to change the power relations between broadcasters and viewers. As Einav (2004) explains: "ITV has the ability to empower the viewer by creating a return path leading to the broadcaster, thus turning the traditional 'one to many' broadcast model into a 'two way' communication process" (p. 6).

Despite expectations, the utopian dream of interactive TV did not materialize during that decade (Van Dijk, and De Vos, 2001). However, we have come closer to it with the technological innovations of the early twenty-first century. Services such as EPG (Electronic Program Guide), VOD (Video on Demand), or DVR (Digital Video Recorder) enable audiences to select what content to watch and when to watch it. Online streaming services such as Hulu, Netflix, or Amazon Prime allow for consumption of televisual and cinematic content whenever and wherever we choose. But given the rapid pace of modern development, technologies and services for interactive TV typically become outdated very quickly (see, for example: Carey and Elton, 2010). Moreover, the introduction of additional digital devices, such as tablets and smartphones have outweighed the importance of developing interactive TV in one device. Instead, today specialized applications and second screen activities enable many of the possibilities initially expected of ITV.

Various services are seen to facilitate a certain degree of interactivity and grant the audience more power and control. When TiVo was introduced in the United States, for example, it promised a revolution in the power relations between viewers and broadcasters by giving viewers control over programming. Andrejevic (2004a), however, revealed the darker side of services of this type, contending:

"The real revolution heralded by TiVo is not the transfer of control to viewers but a leap in the ability of programmers and marketers to monitor consumers" (p. 23).

Thus, although the interactive capabilities of DVR services like TiVo seem to betoken two-way relations between broadcasters and consumers of media, they do not, in fact, create equal relations between the two sides. First and foremost, they produce additional revenues for commercial TV companies, since the audience usually pays for these services, unlike traditional broadcast TV. Even more valuable to the companies is the multifold information they can mine regarding viewers' preferences and choices, all of which are recorded by the interactive services.

SECOND PATH: "ON THE TEXT"

This para-interactive path invites the viewers to make what appears to be an impact on the TV narrative. Discourse on this strategy is generally restricted to voting in popular competition and reality shows, where viewers determine the fate of the participants. Voting for a favorite contestant is a symbolic indication that the producers are answerable to the viewers (Askwith, 2007). Thus, this form of interactivity is associated with the responsiveness of TV to viewers' choices. Audience voting undeniably has commercial significance, which explains why so much professional discourse is devoted to the subject. Nevertheless, the act of voting is only a behavioral manifestation of interaction the viewers are invited to perform. I would like to move the discussion beyond the actual act of voting and outline the different roles viewers can play in relation to the text.

I propose that the para-interactive path that invites the audience to impact the text, even though it is not realized by most viewers, frames the program as more appealing since it offers the individual viewer the opportunity to become a partner, or "coauthor," in creating the show.

Coauthorship of the Narrative

Certain genres, such as reality shows and televisual competitions, lend themselves more readily to the invitation to viewers to impact the show and become coauthors of the narrative. The following examples are all taken from the Israeli version of *Big Brother*. I have chosen to focus on this program for two reasons. First, it is a popular international format known all over the world and still being broadcast successfully in numerous countries. Secondly, and even more importantly, it represents a turning point in the role viewers play in the TV narrative (Tincknell and Ranghuram, 2002). It is they who decide which of the participants will stay or go, thus impacting the storyline and imitating truly interactive narratives. This appearance of interactivity, what I call para-interactivity and the invitation to viewers to impact the narrative is highlighted and even celebrated by the producers as a central feature of the format and an essential element of the program.

In the first example, the host addressed the viewers at home toward the end of an episode and urged them to vote for a certain contestant, dubbed "The Porcupine," who had had an amusing interaction with a cricket the previous night.

> You can't let the "Porcupine" go home today, you just can't. I stayed up all night to see what happened with the cricket. If you send him home today, what will I do at night? (Big Brother, Sept. 2, 2008)

Big Brother is transmitted live 24/7 on a dedicated open channel. Apparently, the cricket incidents had been very popular with the viewers during the wee hours of the night. Thus, by addressing the viewers and encouraging them to vote for this contestant, the host hands over some of the responsibility for steering the storyline to the audience. The viewers can potentially determine not only the outcome of the particular episode, but perhaps more significantly, the possibility of watching future nighttime escapades between "The Porcupine" and the cricket.

Big Brother also offers the viewers more provocative opportunities to serve as coauthors of the developing story. In the same season, one of the main subplots was the slow and hesitant blossoming of romantic relations between two attractive and charismatic participants. As the young lovers were candidates for expulsion from the Big Brother house, viewers were asked to determine not only their individual fates, but also the fate of the budding romance and the direction the televisual narrative would take by voting for them and allowing them to stay together. Over a number of episodes, viewers were given what appeared to be the opportunity to become coauthors of a love story, with their decisions determining the course of the romance.

In another example, the fates of two families were on the line.

> Host 1: Don't separate a father from his daughter. Help Bublil stay.
> Host 2: Don't uproot a twin from his brother. Save Boris.
> (Big Brother, Sept. 2, 2008)

Stressing the viewers' power and responsibility for the participants' destinies, as well as the program's storyline, the host added:

> In such a situation, if this was a different program, we could have performed some kind of manipulation to save them. . . . But it is all in your hands. Only you can save them. (Ibid.)

So apparently the power rests entirely with the viewers. They are not only coauthors, but judges and jury as well, while the production renounces its responsibility and highlights the transparency of the process. Handing over the power to determine the characters' fates, which traditionally lies with the producers, implies a potential role exchange between producers and viewers. The decisions made by the viewers are presented not only as diegetic decisions, but almost as humanitarian verdicts: "Save him." This power is accentuated by stressing the presumably limited power of the production team, which can not pull any tricks or manipulate the results, as well as the helplessness of the contestants themselves who await the audience's decision. It is no wonder that the viewers' omnipotence is a regular issue in the participants' conversations in the Big Brother house, where they express their fears of the way they are perceived by the audience (Holmes, 2008b). It may be assumed that these conversations themselves serve as an empowering strategy for the viewers at home. Their "power" is portrayed as almost mythical, since the decisions they make determine who will have to leave the house and "cease to exist," at least on screen.

In other formats, the audience decides who will be expelled from the tribe (*Survivor*) or who will become a star (*Pop Idol*). This para-interactive path therefore enables viewers to impact not only the televisual text, but also the participants' fate on screen, and implies a dramatic impact on their lives in the real world as well. The winners receive large prizes and the whole course of their careers, and maybe even their lives, may change. All this is presented as a direct result of the choices made by the viewers.

Although the viewers are perceived here as holding all the power, the producers provide them with a select set of criteria and information with which to "judge" the participants and make "informed" decisions. Participants are judged on two main dimensions: objective "professional" criteria ("who is the best"), which relate to the contestants' talents and accomplishments; and subjective "moral" criteria ("who is the most deserving"), which relates to their values, behavior during the show, and "authenticity" on screen. Both dimensions and the relevant criteria are embedded in the texts and are often demonstrated by formal or informal televisual judges. Viewers can choose either to accept or reject their opinions.

Objective "professional" decisions are more suited to talent shows, such as singing, dancing, cooking, designing, or modeling competitions. The viewers at home are invited to become judges and decide which competitor is the best, based on professional criteria. In most cases, there is a professional panel of judges, and so viewers can actually judge both the competitors and the judges. This possibility is sometimes presented as a power struggle with the professional judges (Holmes, 2009). Being able to overrule the decision of the judges is another manifestation of the appearance of changing power relations between producers and viewers, echoing and strengthening the interactive promise of "on the text" para-interactive path.

The second dimension, subjective "moral" decisions, is dominant primarily in programs where the contestants' personalities or more accurately their televisual personalities, are judged, such as *Survivor*. At the end of each episode, a tribe meeting is held in the presence of the "members of the jury," at which the behavior, values, and personality of each contestant are discussed in detail. In the Israeli version, at the end of the season, the viewers at home are also given the opportunity to participate in the final decision after listening to the contestants critique one another.

> [The three finalists sit on small stools alongside the "members of the jury." The host sits opposite them and invites representatives of the jury to address the finalists and explain their decision.]
>
> Jury member 1 [addressing one of the finalists]: The truth is that I am not going to crucify you all like my friends did. I am going to let you all do it

to one another. I want each of you to give me a reason why not to vote for the other finalists.

Finalist 1: I respect each one of them [pointing to the other finalists] and the long way they have come in order to be here. . . . The only thing I find hard is belittling and offending other people.

. . .

Finalist 2: Dan [another of the finalists] is an intelligent person, but in my opinion he played a dirty game. He was patronizing and arrogant and hurt people even when it was not essential strategically. More than me and Noam [the third finalist], who I really love personally. . . . Survivor is a social game, a strategic game, and he was playing some other game.

(*Survivor*, May 17, 2008).

This scene demonstrates the criteria for judging the contestants. Most are subjective, based on affinity, personal qualities, the way they treated the other participants, the way they played the game, and their perceived strategic abilities. The prolonged and recurrent discussions throughout the season spelled out these criteria for the viewers, "coaching" them, as it were, for their role as judges at the season's end. During the live finale episode, Israeli viewers could actually vote for their favorite finalist, with their votes carrying a third of the weight and the rest in the hands of the jury, comprised of contestants previously expelled from the tribe.

Similarly, in *Big Brother* the participants are summoned to the "Diary Room," where they divulge their opinions of their fellow participants "in private." Only later, after a careful editing process, are these testimonials shown to the viewers, allowing the producers to define the relevant criteria for judgment. Here is an example of an edited montage of quotes relating to one housemate who was a candidate for expulsion from the Big Brother house:

Participant 1: Keren . . . the thing about being a hypocrite, it's just disgusting. When she's with me she's so sensitive and two minutes later I find her outside talking about me.

Participant 2: She spreads lies all over the house about almost everybody.

Participant 3: I caught her lying several times. I believe that lying is one of her worst traits and I don't want to see it in the future.

Participant 4: The show she put on in the past two weeks, suddenly she's so sweet to everybody . . . I don't buy it.

Participant 5: I don't have a connection with her and also she's not so good at this game. And that's that.

(Oct. 7, 2008)

The criteria for judgment therefore include considerations such as truthfulness, honesty, relations with other housemates, consistency of behavior, and ability to play the game. Although the viewers are invited to make their own decision, they are given a readymade set of tools to apply. Admittedly, however, they are qualified to pass judgment on the basis of their own life experience, as the Big Brother housemates are ordinary people, just like them (Turner, 2010).

Judging fictional characters in the media is nothing new. Herzog (1941) studied radio listeners and discovered that they derived great satisfaction from critiquing the morality of radio soap characters, and the same was found for the viewers of the TV series *Dallas* (Liebes, 1998). However, in many contemporary televisual texts the characters are ordinary people "playing" themselves, so that the propensity to judge their behavior and personality is even more natural (Hartley, 2008; Hill, 2005; Skeggs et al., 2008). In addition, the judgment can actually be executed by voting, thus impacting the fate of the participants and the narrative of the show.

By enabling viewers to impact the text and determine the fate of its characters, "on the text" para-interactive path implies that the viewers have part ownership of the text and the story. This makes TV producers appear to be responsive to their input in a manner resembling the affordances of interactive TV (Askwith, 2007), and in line with the interactive promise whereby media consumers can become cocreators of media texts (Andrejevic, 2004b).

As we have seen in previous chapters, TV makers perceive the invitation to impact the program as appealing to viewers, believing they wish to feel as if they are part of the show and its production. For this reason, Israeli TV even tends to embed this type of para-interactivity into international formats where they did not originally exist. The international format of *Survivor*, for instance, does not originally include a viewer voting component. However, the Israeli adaptation added a new module: the finale of the taped season is aired live and the audience at home votes for the season winner. Similarly, the original version of *The Beauty and the Geek* does not include an option for viewers to impact the outcome. But in Israel, a live episode added at the end of the season enabled them to determine which of the three couples who made it to the final would win.

It should be noted, however, that interactivity is usually perceived as an individual process, since the user's interaction with the media environment is an individual act based on personal decisions. Thus, the televisual implementation of "on the text" para-interactivity contains an internal paradox. On the one hand, each individual viewer can decide if and how to impact the text, judging and voting for the contestants as they see fit. Yet on the other hand, the behavioral expression of this choice is aggregative or collective, only having an impact when many people vote in the same way. This paradox is symbolic of the status of TV: as a mass medium in a digital interactive mediascape, its use of "on the text" para-interactivity can only produce a pseudo process of "collective interactivity."

THIRD PATH: "IN THE TEXT"

> This rare opportunity to cross the lines and enter into television. . . . The door opens and suddenly you come face to face with Jackie from the TV and Dana from the TV and Frida from the TV. . . . It is a completely different experience. You enter this house and suddenly they all hug you and you feel as if you have known them forever. (Merav Batito, a *Big Brother* participant who joined the program in the middle of the season, Feb. 3, 2011)

"In the text" para-interactive path refers to the various strategies employed by TV to create the impression of an invitation extended to all viewers to enter the televisual text and become participants in it. For decades now, TV can occasionally succeed in making the viewers feel as if they are present at a televised event, such as "media events" for example, even when they stayed at home (Dayan and Katz, 1994), a sense of "presence" that is essential to the viewers sitting on their couch at home so they can feel they as participants (Scannell, 1996). Currently, however, television has been going further than merely producing the sense of presence, offering viewers the chance of a real and actual presence on screen. Thus, TV makers are inviting viewers to cross the line separating the living room from the TV world and enter the "TV sphere" (Holmes, 2009). Before relating to this para-interactive path in greater detail, I would like to chart its history and theoretical roots.

"Crossing the Line": The Viewer Enters the Text

Crossing the line between the real world and the world on screen is a nearly century-old fantasy that was reflected in cinema long before it appeared on TV. In Buster Keaton's 1924 film, *Sherlock Jr.*, the main character jumps into a movie through the screen. Years later, Woody Allen based a whole movie on this fantasy. The heroine of his *Purple Rose of Cairo* (1985) is a cinema buff who steps into the film after passing through the screen in the movie theater she frequents right after at the movie's main hero crosses the line to the real world. Metaphorically, she crosses the line separating the viewers' world from the diegetic world of the cinematic characters. After entering the movie world she does not become one of "them," the performers, but remains a member of the audience who finds herself in the wrong environment. Her crossing of the line is a cause of concern for the fictional movie producers, since they realize it can shatter the cinematic illusion of reality at the basis of any movie. As a result, they rush to restore order and return the main characters where they belong.

In contrast to the heroine of *Purple Rose of Cairo*, who could not fit into the cinematic text, many TV texts today afford the viewers a structured path of entry into the narrative where they can play a central and essential role. The diegetic dissonance which normally occurs when viewers cross what is often called the "fourth wall" between performers and audience (see, for example: Auter and Davis, 1991, p. 165) results from the traditional dramaturgic arrangement (Goffman, 1974) in which there is a clear separation between the stage reserved for the performers and the rest of the theater occupied by the audience. Furthermore, this arrangement presupposes that the audience does not have the right, or the obligation for that matter, to participate in the dramatic activity on stage, except for clapping or cheering the performers. In contrast, the para-interactive path "into the text" appears to invite audience members to become performers themselves. To use Goffman's (ibid.) concept, it suggests that members of the audience can operate within the "performance frame" which positions stage actors in a place and situation that allows other people to watch them intently for an extended period of time without being perceived to violate their privacy. Inviting the viewers into the text transforms them into performers, allowing others to watch them, thus blurring the line between performers and viewers and contradicting Goffman's traditional dramaturgic arrangement.

In recent years we have been seeing more "ordinary people" on screen (Enli, 2009; Lunt, 2004; Turner, 2010). If in the past TV producers set the viewers apart from the show, unless they were members of the audience in the studio, they are now inviting them to cross the line and become part of it (Jermyn and Holmes, 2006, p. 52). It would seem that "ordinary people have never been more desired by or more visible within the media" (Couldry, 2003, p. 102). Some scholars even go so far as to propose that "the viewers have ostensibly taken over the airwaves" (Andrejevic, 2004a, p. 7). This phenomenon echoes Jay Rosen's famous claim that journalism is increasingly relying on citizens' reports, or as he put it: "The people formerly known as the audience."[1] This not only blurs the line between professionals and members of the audience, but also indicates the potential for anyone to become a participant in the media. Jenkins (2006) believes that as part of what he calls "participatory culture," people today claim the right to participate in the production of cultural and media texts. Technological and cultural developments in the digital media environment play an important part in this demand, as individuals can become "instant celebrities" online even if they have no outstanding talent (Turner, 2010). I propose that "in the text" para-interactive path is a way for television to appeal to its viewers in the digital age by appearing to embrace participatory culture.

Strategies for "In the Text" Para-interactive path

Three main strategies are employed to construct the "in the text" para-interactive path:

1. **Direct appeal:** viewers are directly and explicitly invited to participate in the program
2. **Inclusivity**: the invitation to participate is portrayed as inclusive
3. **Representation**: viewers at home are represented by on-screen participants who are just like them.

1. Direct appeal to viewers to enter the text

Many programs today rely on the participation of ordinary people, including reality shows, competitions, game shows, lifestyle programs, self-improvement shows, and others. They are invited to become participants through direct appeals in promos, at the end of an episode, on banners on the TV screen, or online. For example:

> Can you defeat the 100? Register now. Auditions have begun. (promo for the game show *1 vs. 100*, Aug. 1, 2011)

Or:

> For the first time in a reality show and only in Romance Bay you can go straight into the action every week! Register now. (promo for *Romance Bay*, Sept. 4, 2008)

Also:

> The registration to the new season of X FACTOR is now open!
> If you think that you have the X FACTOR, this is your chance. Fill out the form—you might be the big winner in the upcoming of the X FACTOR. Good luck! (Online banner for the 2017 season)[2]

Such invitations may also appear during the program, whether in the form of text on the screen or narration. For instance:

> Do you love to cook? Send us your favorite recipes and you might be invited to participate in the next season of Master Chef. (narration during the 2010 season of *Master Chef*)

Although the ostensible purpose of these invitations is to recruit participants for the show, it is also known that the producers actually search for interesting applicants through a professional search and casting processes. Thus, the direct invitations to the viewers that are embedded in the text are, in fact, used to contextualize the viewing experience within the frame of "participation potential." Indicating the path into the text, they place the viewers at home on the symbolic border between viewing and participating, and frame them as an essential element of the unfolding narrative (White, 2006).

2. Inclusivity

The direct appeal to the viewers is strengthened by visuals and rhetoric aimed at producing the appearance of inclusivity: participation is open to everyone. "It can be anyone" read the text over visuals of a long and crowded line of applicants waiting to audition for the singing talent show *A Star is Born* (promo, May 31, 2008). The camera panned what seemed to be an endless line while the narrator recited a never-ending string of names and places of residence. The whole promo was a multimodal medley of images, sound, and narration, an implied invitation to the viewers at home to become part of the televisual text by sending the message that they could also be there in line if they chose to. To paraphrase Scannell's concept of "for-anyone-as-someone" (2000), which describes the way radio addresses the listeners through a general appeal that appears to be personal, this long line of applicants waiting for the audition constructed what might be called "everyone as someone." In other words, since everyone is invited, it could just as well be you. The crowd of contestants was used as a personal invitation to the viewer still sitting at home in front of the TV. Auditions can be perceived as "dramatization of democratization" (Turner, 2010, p. 19), as the situation represents the promise of a fair chance for everyone. In this promo, the mention of the contestants' place of residence created the appearance of democratization, suggesting that anyone from any geographical location or social status could apply. Another promo for the same show referred to the occasion as "the auditions of Israel" (Aug. 26, 2008), thus raising the invitation and the implied inclusivity to a national level. By emphasizing the inclusivity of the auditions, the producers invite the viewers to come even if they have not registered in advance (Kjus, 2009c, p. 286). The photogenic crowds symbolize both the popularity and power of the show and the inclusivity of the invitation.

Paradoxically, the inclusivity of the invitation also reveals the exclusivity of participation. It would appear that people are encouraged to show up at the auditions only for televisual purposes, that is, so that these mass scenes can be taped and aired to show not only that everyone was invited but that they all came. However, the actual auditions, aired just a few days later, produced a striking contradiction between the appearance of inclusiveness implied by the images of the huge crowds lining up for the auditions and the realization that the large majority of applicants were sent home (Kjus, 2009c). Moreover, programs described as "audition guides" meant to prepare potential applicants, which may also be part of a show's PR content, similarly highlight the built-in dialectics between the inclusivity and exclusivity of the auditions. Such a program was aired just before the beginning of a season of the singing competition *A Star Is Born* (Jan. 26, 2010). Here, two members of the show's panel of judges spelled out a number of guidelines to help applicants pass the auditions, offering tips such as "choose a popular song that shows you in the best way possible," "stand up straight," "be positive," "appear to be confident," etc.[3] The dual message was that even though anyone can audition, few will go on to the next level. Although well-aware of the rigorous selection process, the producers present themselves as desirous of giving everyone a fair chance, and therefore actively helping all applicants to make the best possible impression.

While a so-called audition guide might make some sense in a singing competition, in a show like *Big Brother*, where the contestant's personality and behavior are on trial, it is surprising to see that producers also try and "coach" the viewers on how to become full-fledged participants. Before the airing of the 2009 season of *Big Brother* Israel, a video segment titled "Inside the Auditions" was uploaded to the show's site and subsequently went viral on fans' sites under the heading: "Do you want to be accepted for *Big Brother*? Tips before the auditions for anyone who wants to become the next reality star."[4] In the video, the all-powerful editor of the program discloses to the viewers, as potential applicants, what he is looking for during the casting process.

> Many people are very dominant in the initial "one on one" interview, but I want to make sure they don't suddenly disappear in a group.
>
> . . .
>
> We want to see how creative their thinking is, how people sell themselves, what their inner world looks like.

As a producer I interviewed explained that as a result of this preparation, "people come to the audition ready. They know what kind of type cast they can fill, and even suggest the potential conflicts they can ignite on the show" (Interview with Z., Feb. 20, 2011). Thus, aware of the fierce competition, applicants adapt themselves to the format's requirements.

In short, therefore, the supposed inclusivity of "in the text" para-interactive path, whereby all viewers are invited to "cross the line" and participate in the TV text, also reveals the exclusivity of the process. Clearly, not everyone has the right talent or personality.

Televisual "inclusivity" has evolved over the years in a way best demonstrated by the development of trivia game shows. This veteran genre is a prime example of the "in the text" para-interactive path, since it has often been based on "ordinary people" as contestants. As the following examples illustrate, game shows have become more inclusive over time, gradually allowing ordinary people, without any special abilities or knowledge, to participate.

The first example is the quiz show *Master Mind,* first aired on British television in 1972. As in many other quiz show, the contestants were "ordinary people." However, the challenging questions in selected areas of knowledge required the participation of "experts" or extremely knowledgeable individuals. The contestants were selected following rigorous behind-the-scenes tests both in their areas of expertise and their general knowledge. In each episode, only three exceptional contestants competed, sending a clear message of exclusivity to the viewers at home.

The next milestone in the evolution of game shows was the popular format *Who Wants to Be a Millionaire?* (1998), its name implying a direct and inclusive invitation to the viewers to participate. In each episode there were ten participants who were chosen on the basis of pre-show general knowledge tests. The innovation in this format was that the final round of "tests" was conducted in the studio as part of the broadcast, so that viewers could see which of the ten contestants got the chance to "become a millionaire." The fact that the contestants were "ordinary" people and not necessarily experts made this format seem more inclusive than previous quiz shows.

A more recent format, called *1 vs. 100*, originally aired in the Netherlands in the year 2000 and was later distributed globally. Here, a single contestant competes against 100 people seated in a small-scale arena. The constellation of the studio blurs the line between audience members and performers. Visual dissonance is created, as the large number of people and their sitting arrangement makes them appear to be the audience, but they actually play an active role as participants in the program. I believe this dissonance was intentionally designed to symbolize the inclusivity of "in the text" para-interactive path.

The last game show I would like to mention in this context is the British format *Cash Cab* (2005), which has been adapted and broadcast in many countries and languages. Conducted in a cab driving around the streets of the city, the driver serves as the host of the show and the contestants are the unsuspecting riders who hail the cab in order to get to their destination. In the long line of quiz programs, this format seems to be the most inclusive. Ordinary people on an ordinary day hop in a cab and, without any audition, preparation, or even prior notice, are transformed into participants in a TV show. They are not required to have any particular talent or special knowledge, or even to express their desire to participate. It seems as if without any warning whatsoever, anyone can become a contestant in this quiz show. Thus, *Cash Cab* is as inclusive as it gets, or at least it appears to be, since it is more than likely that only "interesting" participants survive the editing process and make it to the screen. Nonetheless, the message to the viewers at home is that they have as good a chance as anyone else of crossing the line into the televisual text. All they have to do is hail a cab.

Quiz shows like these encourage viewers to "jump into" the TV (Holmes, 2009). Since they appear to set almost no prerequisites as far as potential participants are concerned, the invitation to do so appears to be inclusive. As Kjus (2009c) puts it: "Since everyone is a potential participant, everyone gets firsthand participation experience" (p. 283). In effect, therefore, creating an appearance of inclusivity is mainly a way to appeal to those who do not accept the invitation to "jump in."

3. Representation or participation by "proxy"

Another "in the text" para-interactive strategy meant to produce "a firsthand participation experience" is to create the impression for viewers still in their living rooms that they are represented on screen by people who closely resemble them.

> Nowadays you can not only live with people like you, eat with people like you, and complain with people like you, but you can also watch people like you on TV. (Smadar Shiloni, "No Escape," YNET, March 10, 2010)[5]

This strategy is aimed at enhancing the feeling of "presence" in the televisual text by participation "by proxy."

> Everyone knows that even if they are not there, they do get some kind of representation in the program. (Ibid.)

In my interviews with TV producers, they have explained what they look for in the casting process in order to create this impression of representation.

> [The cast should be] mainstream but interesting . . . anyone should be able to find himself in it. (Interview with P., Dec. 25, 2008)

Along the same lines, one of the most popular winners of *Big Brother* in Israel was depicted in the media as a generic persona, in terms such as, "here comes Eliraz Sade. . . . There are thousands like him." Such descriptions might explain the popularity of this contestant. In the eyes of the viewers, he was just an ordinary guy like thousands of others, representing them, their sons, or the next-door neighbor. To borrow from Scannell (2000), we could say that Eliraz was the human epitome of "someone-as-anyone," someone that anyone watching could see himself in.

Representation of the viewers by people like them was traditionally created by having a live studio audience (Horton and Wohl, 1956). Its presence was originally intended to demonstrate to the viewers at home the "right" response to the show, such as when to laugh or clap. At the same time, however, the studio audience symbolizes the presence of the viewers in the text (Peters, 1999), serving as a "mental bridge" and representing them within the program (Kjus, 2009c, p. 287). While this form of representation is a televisual tradition, in recent years "ordinary people" have been participating in the text not only as audience, but also as performers and main characters. Consequently, the viewers at home can now see their "lookalikes" actually performing on-screen (Andrejevic, 2002; Coleman, 2006; Kjus, 2009b; Palmer, 2002).

Media texts have always offered readers or viewers characters with whom they can identify (see, for example: Friedberg, 1990). However, TV makers I interviewed believe that currently TV affords more than that.

> TV today tries to offer more than it used to. It still offers identification but more than ever it offers an opportunity [to participate]. . . . Most people won't come to the show, but they see themselves as [the participants] since even if you're not really there it's almost as if it's you on screen. (Interview with D., Dec. 25, 2008)

In some cases, viewers are made to feel almost as if the characters are their own "avatars" whom they can operate and impact as they choose. As eloquently described by a journalist who met the winner of reality show for an interview:

> For her I am a complete stranger, but for me she is my elegant Barbie doll, the doll I played with and dressed and moved and activated for three months. (Spector, 2008, p. 8)

Or in the words of a TV producer:

> We wait to see how these people will behave for us. . . . We see them as our extensions. (Interview with S., Dec. 14, 2009)

Thus, the participants on screen can be perceived as representations or "extensions" of the viewers whom they can "activate" and whose moves in the game they can determine, much like avatars in a virtual interactive environment. TV makers therefore seek to appeal to viewers by sending the message: "You are invited into the text, but even if you prefer to remain at home in front of the TV, you can still participate in the program by proxy through your 'avatars' in the text."

To conclude, "in the text" para-interactive path invites the viewers to become part of the televisual text by employing three interconnected strategies: direct appeal, inclusivity, and representation. If here TV attempts to mimic the digital online environment, the next para-interactive path consists of actual online interactions that relate to the televisual text.

FOURTH PATH: "ALONGSIDE THE TEXT"

This path of para-interactivity allows viewers to interact and engage with the program, the production team, or other viewers through online platforms and social media during or after the airing of the show. Online interaction is referred to in the television industry as "second screen transmedia," "cross platform TV," or "social TV." As the subject has been discussed extensively in both industry and academic forums, I will relate to it here only briefly.

Online platforms can serve a number of purposes. In some cases, they offer viewers more information on a broadcast, such as sports statistics (Sørensen, 2015) or the detailed recipes for dishes shown in cooking shows. In shows like trivia quiz programs, viewers can sometimes play along during the broadcast, participating online using their own devices at home. For example, the website for the Israeli version of *The Million Pound Drop*, greeted viewers with the message:

> Welcome to the show, the first TV game where you can play at home along with the show, competing against the contestants. You can even win real money, a lot of money. So prepare your laptop in front of the TV and join in the game.[6]

Social media platforms enable viewers to read posts or tweets from the production team or to post their own comments and suggestions for the program, addressed to the producers. They can also converse online with other fans, tweeting their impressions and criticism. By urging viewers to engage with the program on social media, TV makers not only create the impression that the people at home are participating in the show, but also, and perhaps mainly, generate a buzz around the program and earn free promotion (Keinonen and Klein Shagrir, 2017) and as studies shows, simultaneous use of second screen alongside television, enhances engagement of users

(Nee and Dozier, 2017). Engagement on social media is open to all viewers, and appeals even to those who do not choose to interact in any other manner with TV shows, whether in the text, on the text, or alongside the text.

FIFTH PATH: "IN FRONT OF THE TEXT"—"INTRA-ACTIVITY"

> When you watch these ordinary people within a big thing [a TV show], it makes you wonder how you would behave if you were there. (Interview with P., Dec. 25, 2008)

The concept of intra-activity draws on Lotman's (1990) typology of human internal communication, that is, within the self, which differs from the individual's interaction with another person. According to Lotman:

> The "I-s/he" system allows one merely to transmit a constant quantity of information, whereas the "I-I" system qualitatively transforms the information, and this leads to a restructuring of the actual "I" itself. . . . While communicating with him/herself, the addresser inwardly reconstructs his/her essence. The transmission of a message along the "I-I" channel is not a self-contained process since it is caused by the intrusion of supplementary codes from outside and by external stimuli which alter the contextual situation. (p. 22)

Lotman highlights the change or development in the self that is the result of inner communication, a discourse that is not closed, but is also impacted by external norms and messages which enable the person to examine and reconstruct his or her own self. In line with Handelman (2003), who ties this concept to television, I propose that the televisual text can serve as an external trigger to an internal interaction that encourages the viewers to examine themselves and their lives, and may even induce a process of transformation. Handelman describes the meeting between the viewer and the TV screen as a virtual encounter, "one that triggers interaction among the voices that constitute the selfness of the viewer" (2003, p. 147). While his focus is on the medium and the screen, I suggest widening the scope of this claim to include the televisual text and the viewers' encounters with it. Since many current televisual texts contain "ordinary people" and narratives that are relevant to the viewers' potential participation, they can readily "trigger interaction" among the voices within the viewers' selves and thus invite them to engage in intra-activity "in front of the text." In other words, contemporary televisual texts are likely to cause viewers to examine their self, their abilities, and their talents, as well as their norms and behaviors, in light of the preconstructed criteria presented in the program and by comparison to the ordinary people participating in the text. Since intra-activity is actually an invitation to impact the self in front of the televisual text, it may be thought of as "reverse interactivity": viewers do not impact the text but are impacted by it.

The intra-active path can lead the viewers in three potential and interconnected directions:

1. Viewers can examine how they would conduct themselves if they were participants in the program.
2. Viewers can rate themselves and their lives in relation to the participants in the program.
3. Viewers can learn and improve themselves as a result of intra-active process in front of the text.

Whereas the first direction relates closely to the TV world, the other two relate to the real world, the viewers' everyday lives. As I explain below, sometimes they overlap, and all three may even be found in the same text.

Viewers Can Examine How They Would Conduct Themselves If They Were Participants

Viewers who go in this direction imagine themselves within the text, in the shoes of the participants, and ask themselves questions such as: Would I be willing to participate in such a show? If I was there, how would I behave? How would the viewers at home perceive me and my behavior? Could I survive the situation? One well-known Israeli TV critic formulated this situation in a highly cynical fashion, claiming:

> Many Israelis—well not many—all Israelis—have already asked themselves, more than once, how they would have survived the Holocaust if they'd been there. . . . The question Israelis ponder these days is how they would survive a TV show. (Shaked, 2009, p. 42)

"In the text" para-interactive path, discussed above, promotes the inclination for intra-activity because the participants in the televisual texts are ordinary people. Since the viewers at home believe they could potentially be one of them, the inner discussion is not merely theoretical. One producer explained the appeal of such programs.

> You fantasize about being there. . . . It's not a famous star in there, it's a person like me . . . I could be there tomorrow or in a year or two. So let's see what kind of decisions he has to make. He's one of us. (Interview with B., Aug. 26, 2010)

In an interesting twist, one former viewer of *Survivor* who ultimately became a contestant described his inner debate when he was still watching the show from his couch at home.

> It was great to eventually play the games that you see on TV, when you say to yourself: Wow, look how he does it, Why did he do that . . . I could do it better. And suddenly you are really there in a challenge.[7]

This quote not only illustrates the viewer's inner discussion in front of the text, but also links it with what appears to be the realistic potential to enter the text and become a participant. Indeed, in this case the potential was actually realized.

Viewers Can Rate Themselves and Their Lives

Another direction the intra-active path can take is for viewers to rate themselves according to a scale built into the program, comparing their achievements to those of the participants in the show. This is characteristic of programs such as quiz shows, where viewers can judge their level of knowledge as compared to the contestants on screen. One TV maker described the appeal of the game show *1 vs. 100*.

> We want the viewer to sit in front of the questions and try to see if he is better than the person who was chosen to be a contestant, to ask himself what he would do in his place, take the money and go home or take the chance and go on to the next level. . . . I sit at home and ask myself what do I do? Do I know the answers and stay till the end or not? . . . The viewer at home can say to himself, tomorrow I can be one of the 100, and let's say I am, how much money will I take home? They can actually compete against the contestant there on stage if they choose to, and if they don't know the answer they want to see how many of the 100 didn't know the answer either and dropped out with them. (Interview with K., Sept. 7, 2009)

Along similar line another quiz show integrates the invitation to the viewers in the title of the program: *Are you Smarter than a fifth grader*. TV producers thus appear to understand and acknowledge the attraction of self-rating for the viewers. However, while in quiz shows the rating scale is clear and easily followed, in other programs viewers are invited to examine themselves and their achievements in everyday activities according to a vaguer set of criteria. Female viewers, for example, can rate their own mothering skills by comparing themselves to the mothers on *Wife Swap* (Hill, 2005). This type of rating is allied with the third direction intra-activity can take: self-improvement.

Viewers Can Learn and Improve Themselves

In self-improvement and lifestyle programming, the invitation to the viewers to learn and improve is direct and overt. Indeed, these programs are often perceived as "learning opportunities" (Hill, 2005). For example, in one episode of *Overdraft Family*, which helps families with financial and life challenges, the coach looked straight at the camera and invited the viewers to examine their own situation and learn from what they have seen on screen.

> From the Hamiel family, from the process they have undergone, you can learn something very important. . . . When it is about us . . . we can't usually help ourselves [images of the family arguing in the background]. So my advice is to know when it's time to go out and consult someone else. . . . It can be a friend, a professional, your mother or father, I don't know [v.o. images of the coach taking a blindfold off the mother of the Hamiel family. He hugs her while she cries. . . . Cut to the coach, again facing the camera], someone who is acceptable to all sides so he can help us look at ourselves from the outside.[8] (*Overdraft Family*, season 5, Ep. 2)

Thus the coach/host addresses the viewers directly, inviting them to look hard at themselves, their lives, their relationships, and especially their financial conduct. He offers a televisual mirror in which people can see themselves as they watch families like them with similar or worse problems being coached to better their lives. In front of this text, viewers can choose to take the para-interactive path of intra-activity, to "take the blindfold off," examine their own lives, and conduct an inner discourse triggered by the family on screen. Moreover, when the TV coach urges viewers to reach out to "someone who is acceptable to all sides so he can help us look at ourselves from the outside," he seems to be offering television, and his own program in particular, as a viable option. In fact, the program is explicitly presented as a "learning opportunity" in the promo, which states: "You should watch this program for your family" and "Every family has a story that can be changed" (Dec. 27, 2011).

It is not only self-improvement shows per se that invite the viewers to transform themselves. Even programs like *Big Brother* are often described by the producers as endowing the viewers with tools for life (Andrejevic, 2006). Here the invitation to learn and develop "in front of the text" is more covert and is issued primarily through the discourse of the participants, who often attest to undergoing a process of self growth and getting to know themselves better by being on the show. This rhetoric persists in their discourse even after the season ends (Hill, 2002), when they continue to relate in interviews how much they learned about their inner self (Palmer, 2002). They describe a process of learning, of becoming true to their real self, etc. (Andrejevic, 2004a, p. 108), thus portraying the programs as guides for "ways to live" (Hawkins, 2001).

An interview with one of the Israeli winners of *Big Brother* illustrates this framing of the show.

> Interviewer: What made you go for it?

> Eliraz: Everything. The prize, to be famous, future opportunities. Everybody wants it and anyone who doesn't admit it is a liar. In the end, I benefited much more. I learned a lot about myself and I think I showed the people out there the right way to live. (Abramovitz, 2010, p. 15)

Eliraz describes how he took the "in the text" para-interactive path and became a participant in the program. He candidly admits he wanted to become famous and stresses that "everybody wants it," thereby presenting the path as a valid option for the viewers at home. In addition, he claims that as he was transforming himself during his time on the show, he was also helping the viewers to go through a process of personal development and improvement "in front of the text" by showing "the people out there the right way to live."

Thus, the different directions in which the intra-active path may go often overlap and intertwine. Viewers can consider how they would behave if they were to participate in the program, rate themselves in relation to the participants, and try to improve themselves accordingly—all at the same time and in front of the same televisual texts. Presenting programs as guides to self-examination and self-improvement is therefore another para-interactive strategy employed by TV makers today to attract viewers.

CONCLUSION

This chapter defines five para-interactive paths adopted by contemporary television which invite the viewers at home to engage in a variety of actions and interactions, either real or potential. **In front of the Screen** para-interactive path affords the viewers control over what televisual content to consume and when and where they watch it; **in the text** para-interactive path invites them to become participants in the text; **on the text** para-interactive path offers them the opportunity to impact the text by becoming coauthors or editors; **alongside the text** para-interactive path enables participation in show-related activities through a second screen or interaction with the producers and other viewers through social media; and **in front of the text** para-interactive path strives to generate intra-activity whereby viewers can examine, rate, and even improve themselves by seeing people like them on screen. Thus, viewers have a range of paths to choose from, and their choices shape their viewing experience. They can take one path or more, or even no path at all, echoing the options in Borges's "Garden of Forking Paths."

McLuhan claims that new media attempts to resemble older media, at least until they succeed in developing their own unique characteristics and can break free of the past (Scannell, 2000). What I have described here might be seen as the same phenomenon moving in the opposite direction: the strategies adopted by an older media, television, in an effort to resemble new media. By inserting various types of para-interactivity into televisual texts, TV is seeking to mimic interactive digital media in order to maintain its relevance in a changing mediascape. In this manner, TV is striving to appear responsive to viewers' input even though it is basically a one-way medium. The para-interactive strategies it employs are inspired by the interactive promise of participation and impact of media consumers on media messages (Andrejevic, 2004a). These options, however, offer only limited and strictly defined opportunities for participation and impact. To date, at least, broadcast TV can not allow for the same sort of interactivity afforded by digital media, and consequently power relations between TV producers and viewers have not changed dramatically.

NOTES

1. "The People Formerly Known as the Audience," Rosen http://archive.pressthink.org/2006/06/27/ppl_frmr.html (accessed: Feb. 25, 2011).
2. Reshet online site: https://www.shortaudition.com/X-Factor (accessed: March 15 2017).
3. This segment was later uploaded to the broadcaster's online site: http://www.mako.co.il/music-24/24new-yasu/Article-9090768334b6621006.htm (accessed: 10.10.2016).
4. http://www.tapuz.co.il/blog/ViewEntry.asp?EntryId=1598482&passok=yes&r=1 (accessed: Feb. 25, 2011).
5. http://www.ynet.co.il/articles/0,7340,L-3860185,00.html (accessed: Dec. 18 2016).
6. http://www.mako.co.il/tv-million (accessed: Sept. 8, 2011).
7. "Who's the King of the Island? Interview with the Stars of *Survivor*," *Pnai Plus*, June 1, 2010 http://www.ynet.co.il/articles/0,7340,L-3897241,00.html (accessed: June 5, 2010).
8. *Overdraft Family*, Reshet, Channel 2, Season 5, Episode 12. http://reshet.ynet.co.il/Shows/Mishpaha_Horeget/videomarklist,183207 (accessed: Sept. 12, 2011).

Chapter Four

"Live" TV as a Para-Interactive Temporal Strategy

Many TV makers believe that live programming is what broadcast television needs today, as they openly admit their concerns regarding television's future in the digital media environment:

> Everything that's happening now in "the digital" proves that people want live broadcasts. Young people watch TV while they're on the iPad or iPhone, posting and tweeting—that's a very strong trend. People want to be "in the moment." That's why sports work so well. TV needs that kind of content that has its "moment in time." In other words, if you missed the broadcast . . . you're not part of the conversation. . . . People want to be the reporters . . . that's Facebook, Twitter, and Instagram, so I truly believe that live—live events and live programming—that's where the future is going.[1] (Assaf Blecher, Keshet Broadcasting [Channel 2, Israel], formerly V.P. of programming and development for Dick Clark Productions [USA])

In their eyes, live TV is the remedy for television's challenges such as the diminishing linear viewership and the erosion of the commercial TV model. Moreover, they attribute the decline in revenues from advertisers to young audiences, who can watch audiovisual content whenever and wherever they choose.

The opening quote touches on a number of my arguments in this chapter, which relates to live television in the digital age. As Mr. Blecher points out, not only do live broadcasts attract viewers to the linear television schedule, but they also invite them to be, as he puts it, "in the moment" and feel part of what is taking place on screen. Moreover, they enable people to engage with the program and be "part of the [online] conversation" about it at the very time of its transmission. Thus, the "liveness" of television can extend beyond the broadcast itself by means of digital platforms and social media that serve as "second screen" activities.

Is live TV indeed the future of television? This claim is intriguing and deserves further examination for two reasons. Firstly, whether or not it is the future of TV, it is definitely a central element of its past legacy, and, as I will demonstrate below, it is already a major part of its present. Secondly, the claim that live TV is the future seems to contradict the fact that television content is consumed today in different temporalities at the discretion of the viewers, and therefore detached from TV schedules. Nevertheless, broadcast television is being pushed to produce programs that are best watched live in order to maintain the viewership of linear broadcasting and the advertising revenues it brings with it.

I argue that a "live" viewing experience has become even more significant in the changing media environment, as TV makers perceive it to be part of the appeal of TV in the digital age. However, "live" TV today comes in various shapes and temporalities, and carries new meanings and purposes in addition to the traditional ones. I conceptualize the desire of television producers to create a live viewing experience as a temporal para-interactive strategy that has the potential to strengthen para-interactive paths for participating in and impacting on televisual texts as well as for engaging in online and second screen activities along the televisual text (detailed in chapter 3 of this book). Furthermore, I claim that television produces "seemingly live" programs that is, taped non-live shows infused with "liveness indicators."

Below I offer a brief review of the concepts of "live" and "liveness" in relation to television, and sketch the trajectory of their history. I then discuss the various strategies and means employed by contemporary broadcast television to create an appearance of "live television" through "liveness indicators," even when the shows concerned are taped, and offer new modes of liveness: "second shift liveness" and "alternative liveness" produced through online platforms and social media. Finally, I contextualize live programming within contemporary purposes and meanings in the digital mediascape. I propose that television strives to link viewers not only to real world events, but also to the televisual text, its production, and its participants in an ongoing temporal connection. Thus, I relate here to live TV as a temporal para-interactive strategy that helps television strengthen its connection with viewers, while enhancing and validating the para-interactive paths discussed in the previous chapter.

WHAT IS LIVE?
THE CONCEPTS OF LIVE AND LIVENESS ON TELEVISION

The concept of a live broadcast basically refers to the synchronization between the time of the televised event, its transmission, and its viewing (Feuer, 1983, p. 14), or to the lack of an "interval between the production, transmission and reception of programmes" (Scannell, 1996, p. 105). Although Bourdon (2000) offers a more detailed typology of segments within televisual content, defining various types of "live," his definition of a "fully live" broadcast similarly relates to the time of the event being parallel to the time of the transmission and viewing. From an ontological and technological perspective, a broadcast "comes to you live" when the various elements of its production—the event, filming, editing, transmission, and viewing—all take place simultaneously. Media events are the prime example of "fully live" broadcasts: big events that call for live viewing are broadcast simultaneously, producing a shared experience for the viewers (Dayan and Katz, 1994).

"Liveness" is a related concept. Although it may overlap with "live," it does not presuppose that the program is actually transmitted live (Bolin, 2009). "Liveness" often signifies rhetorical and other televisual indicators of the "here and now" (Ellis, 2000; Tolson, 2006).Thus, whereas "live" emphasizes the technological and ontological, "liveness" refers to the televisual aesthetics and rhetorical characteristics of television that aim to produce a simultaneous and immediate experience. The two concepts are used below by TV makers, interchangeably, depending on the context.

With the introduction of digital media (when television became the "older" medium), liveness was considered one of the major features of the new media (Couldry, 2003). Initially, two forms of "digital media liveness" were identified: "online liveness," which is created by the affordances of the Internet; and "group liveness," which represents the ability of friends to interact via their mobile phones (Couldry, 2004). This distinction has become blurred ever since the arrival of smartphones, where communications are also conducted using the Internet. Digital media technology was originally characterized as having "real time ontology," which enables the retrieval of data and content on demand (Manovich, 2001). This definition was also adopted to distinguish between live broadcast television that transmits events in real time, and "interactive TV" that allows for a real-time response to the user's actions (Bennet, 2008). The distinction between the two meanings of "real time"; the televisual versus the digital is clear and dictated by the different affordances of the two types of media. I contend, however, that "live" on TV is a temporal para-interactive strategy designed to imitate digital "real time." Digital "real time" marks the ability of users to interact, participate in, and impact the online environment and content. I maintain that live television is similarly an appeal to viewers to interact with the televisual texts through para-interactive paths.

LIVE BEGINNINGS: A BRIEF HISTORY OF LIVE TV

Live broadcasting is older than television itself. It can be traced back to the radio, which offered live dramas, musical programs, and news, as well as listeners' phone-in programs, all broadcast live (see: Scannel, 2003). Television was born later as a live medium: in the 1930s and 1940s, all programs were transmitted live. The concept of "live television" was only introduced following the invention of recording devices, when it became necessary to distinguish between programs that were transmitted live and those that were prerecorded (Auslander, 1999). Yet even though live broadcasting was initially the result of technological limitations, television producers celebrated it as a unique feature of the televisual medium (Caldwell, 1995). In many ways, it came to be considered the essence of television and its aesthetics as a medium (Auslander, 2008, pp. 12–13; Corner, 1999; Ellis, 1982; Levine, 2008, pp. 394–95; Marriott, 2007). In contrast to film, where the cinematic image is seemingly a documentation of past events frozen within the frame, television was held to be based on the process, on motion, on the immediacy and simultaneity of the live transmission (Feuer, 1983).

Historically speaking, the early 1950s in the United States and Europe was the golden age of live television (Bourdon, 2000). American networks, for example, broadcast many hours of live dramas during prime time, garnering high ratings (Hawes, 2001). Live broadcasts gradually disappeared from the screen in the late 1950s and early 1960s, mainly for organizational and economic reasons (Bourdon, 2000). Some claim that the continued decline in live broadcasts was the result of technological developments, such as the home video (Cubitt, 1991), as well as aesthetic changes (Caldwell, 1995). Ellis (2002) even states categorically: "The history of television genres is the history of the development away from live performance and transmission to pre-recording of programmes" (p. 33).

In light of the claim that live TV is the future, Ellis's conclusion seems remarkable, although it was accurate at the time. For many years, live broadcasts dwindled, until they became unique to newscasts and special events (Couldry, 2004), while most of the other programs were taped. Nevertheless, television transmission itself was always considered live insofar as it was watched simultaneously by millions of people, and the concept of "flow" became synonymous with the immediacy and presence of TV (Williams, 1974).

Thus, for many years, even in taped programs, when only the last two links in the broadcast chain (transmission and viewing) occurred simultaneously, television was considered a live medium as the viewing experience could be shared temporally with other viewers (Ellis, 2002). McLuhan described this as "all-at-oneness," the bridging of geographical distances by the media through the consumption of the same content by many people at the same time (Meyrowitz, 2003).The television viewing experience was therefore perceived as live since it existed only during the time of transmission. In the words of Cubitt (1991):

> The broadcast flow is . . . a vanishing, a constant disappearing of what has just been shown. . . . Television's presence to the viewer is subject to constant flux: it is only intermittently "present" as a kind of writing on the glass . . . caught in a dialectic of constant becoming and constant fading. (pp. 30–31)

The nature of the viewing experience in television's live transmission is described vividly in the following account, in which the writer reminisces on the fleeting and shared experience of watching the news in Israel in the 1970s:

> In every house on the block, fathers would get ready to watch the evening news. The phone would not ring, no door would be knocked on, and no bird would tweet during the news broadcast. There were still no video recorders, and every missed word was considered lost forever, every smile or raising of a brow was gone, never to return. Every moment during the broadcast was perceived as historic. . . . This hallowed viewing was not unique to the newscasts, but even in other shows I had to watch in silent attention. (Livne, 2009)

This humorous description reminds us that live television is not merely a technological phenomenon. It is also a "conventional expectation," an assumption of simultaneity and shared viewing (Couldry, 2002, p. 286). In other words, the viewers' cooperation is essential in producing a "live" experience, which can be obtained only in a certain environment in both technological and cultural terms.

The traditional link between television transmission and viewing began to unravel with the invention of VCRs (Cubitt, 1991). The trend continued with the introduction of DVD box sets, Pay-Per-View TV, DVRs, and VOD services, all of which have led to "media time squeezing" and the privatization of viewing (Moshe, 2012).

IS LIVE TV DYING?

The technological innovations mentioned above have changed television viewing and the cultural practices surrounding it, while arousing deep concerns in the television industry and even predictions of the death of broadcast TV. In the words of Micha Reis, a TV channel branding specialist: "Within five years the TV screen will be completely different than it is today. Very soon when we turn on the TV there will be no broadcast but a menu to order programs. . . . TV schedules will not exist" (*The Marker*, Aug. 3, 2009).

The fears of broadcast TV executives, who rely on the thirty-second advertising spot and rating-based revenues, are understandable and well-founded. Indeed, the industry discourse regarding live broadcasts' imminent death is echoed by academics, who have claimed that live broadcasting is no longer a valid feature of television (Uricchio, 2004).

Nevertheless, the grim prophecy of broadcast TV's disappearance has not yet materialized. Despite the industry's anxieties, viewers have not deserted linear transmission in favor of personalized viewing schedules and content all together. In fact, TV is more live than ever, and broadcast TV's survival seems to depend on the continued viability of this characteristic. As one TV executive put it:

> It's the result of the "ad busters." . . . We want you in front of the screen now. . . . Live broadcasts offer a solution to TV's need to draw viewers . . . a critical mass of viewers to a certain point in time. Now! Don't record and watch it later, watch it now! (Interview with O., Sept. 7, 2009)

Understandably, then, broadcast TV is trying to create "must-see-now" programs, and invests a lot of effort into inviting the audience to share in the synchronous viewing of linear TV.As indicated in the quote at the start of this chapter, television executives consider live television a major attraction, even a "must," in today's changing media environment.

"MUST-SEE-NOW" TV: HOW LIVE AND "SEEMINGLY LIVE" PROGRAMMING IS PRODUCED

Below I elaborate on the strategies employed by television to create live TV, and "seemingly live" which is taped programs that appear to be live. I begin with my findings from the analysis of Israeli television schedules (2008–2015) that confirm the centrality of live programming, and then present a detailed content analysis of programs, paratexts, and additional texts from online platforms employed by television channels. I will show how all these means are used to produce an invitation to a "must-see-now" viewing experience of live programs, as well as taped programs that strive to appear live.

Prime-time Television is (Mostly) Live

In light of the claim that live TV is the future of TV, it was somewhat surprising to discover that it is already the present. A detailed quantitative analysis of the programming schedules of Channel 2, the most popular TV channel in Israel, revealed that between 2008 and 2015, the majority of prime time (18:00–23:00) programs were broadcast live. More precisely, around 60 percent were tagged "live,"[2] although some were not "fully live" according to Bourdon's (2000) definition, but contained some taped or edited segments. The contextual framing of the programs by the broadcaster, however, was live, and most importantly, they were clearly marked as live by the channel itself.

It should be noted that any study of Israeli TV must take into account the numerous current affairs programs that occupy a substantial portion of the schedule. Not surprisingly, Israelis are avid news consumers, and in times of a security crisis, television becomes a central source of news for the Israeli public (Liebes, 1998). During the conflict in Gaza in the summer of 2014, for example, all three Israeli broadcast channels were live on air almost eighteen hours a day, and drew large audiences continually. My analysis, therefore, did not include these unique times of live broadcast marathons. I analyzed only the planned TV schedules, without including last-minute changes due to emergency situations. This makes the finding of 60 percent live programming time even more striking, since in reality there is always the possibility of additional unplanned live broadcasts.

In addition to the news, a large portion of live broadcasting is taken up by the numerous reality/competition shows that inhabit prime-time television in Israel (Neiger, 2012). The local versions of globally popular formats, such as *Pop Idol* or *Big Brother*, attract the desired linear live viewing and high ratings.

Some genres, such as news and sports, will probably always be watched live. That said, it should be remembered that live programming is not necessarily genre dependent. To give two examples: In the U.S., the drama series *E.R.* broadcast the first episode of the 1998 season live, and even did so twice so that it would appear in prime time in the various time zones (Levine, 2008; Sconce, 2004); In Israel, the popular satiric program *Eretz Nehederet* (akin to *SNL*) broadcast several episodes live during the seasons of 2008 and 2010. The producers explained that the decision was based on the viewers' desire to experience the "energies of the live broadcast" and "[make the viewers] wait with baited breath to see if things go wrong during the broadcast" (Interview with M., Dec. 14, 2009).

At the start of the millennium, Bourdon (2000) contested Caldwell's (1995) claim that television shies away from live broadcasting, contending that it is still employed extensively throughout the world (p. 533). My analysis of prime-time Israeli commercial television supports Bourdon's contention. Not only has live television refused to go away (Marriott, 2007, p. 59), but it has become a central feature of broadcast television in the digital age (Sørensen, 2016b).

Broadcasting live definitely answers television's ambition to appeal to viewers with "must-see-now" TV. Moreover, a detailed textual analysis of the programs and accompanying texts, such as "paratexts" (Gray, 2010) and online content, reveals that producers infuse even non-live, taped programs with "liveness indicators" in order to make the show appear live.

Paratexts as Liveness Indicators

Paratexts are additional texts that accompany and relate to TV programs, including promos, public relations texts, and other media coverage (Gray, 2010). Since paratexts can determine the context of the viewing experience, they can also serve as liveness indicators, injecting a feeling of "live" into taped programs so that they appear to be live.

By way of example, I will focus on two international formats that have been adapted locally, and shown on prime time TV in Israel: *Survivor* (broadcast in Israel since 2008) and *Paradise Hotel*, an American format titled *Romance Bay* in Israel that lasted on air only one season (2008). Both shows are taped and edited ahead of their transmission. However, the producers added a number of live segments: each season of *Survivor* ends with a two-hour live finale episode; and the taped episodes of *Romance Bay* were peppered with live segments. Moreover, since the shows were basically not live, the producers tried to construct the appearance of liveness and presentness by carefully orchestrating the paratextual discourse around them, using media coverage, promos, and ads that functioned as liveness indicators.

As one producer I interviewed explained, media discourse surrounding *Survivor*, or more accurately, the lack of it before the program is aired, helps to create the impression of temporal synchronicity between the televised events and the time of the broadcast. Media coverage is tightly controlled to serve the program's interests.

> All the participants sign a [confidentiality] agreement so they can't disclose any details or give interviews . . . just think about it, how do we know what is happening now? Newspapers, media, paparazzi . . . we make sure all this stuff is cleaned up. The participants have been back from location for a long time, but nobody talks about it. (Interview with O., Sept. 7, 2009)

In contrast to the absence of media reports before the broadcast, a great deal of effort is invested in generating lively media discourse when the program is finally aired. This is also carefully planned: interviews with *Survivor* contestants appear only after they have been voted off, have presumably returned from the distant location and are back in the "real world." These interviews are typically conducted in the present tense, with the contestants being asked how they feel about their departure from the program as if it actually happened the night before, when the relevant episode was broadcast. For instance:

> Abigail Pearl knows exactly who is to blame for her expulsion yesterday from Survivor and she says: I don't want to meet Talko and Shai ever again. They're not my cup of tea. (Porat, 2009)[3]

The presence of this participant in the "real world" outside the diagetic sphere of the show, along with the absence of the other participants whose expulsion has yet to be broadcast and therefore are presumably still on location, creates a temporal link between the viewers' temporality and the narrative temporality of the program and thus implies the appearance of liveness. In the same vein, a video showing one of the participants in a similar format titled *Mission: The Amazon* (2016) on her return home from the distant location was released the day after her departure from the program was broadcasted, even though the episodes were taped and edited more than a year before they were aired.

> We escorted Lee Levi from the airport to her home in Tel Aviv in order to hear from her how she sums up the whole experience. . . . A first interview. (Mako, Sept. 9, 2016)[4]

This video is presented as if it was made immediately after Lee was forced to leave the competition because of an injury. The fact that it is tagged a "first" interview is ironic not only because the incident occurred a year earlier, but also because the contestant herself is a local celebrity who often appears in the entertainment media. Nevertheless, it was framed as recent and exclusive in order to imitate "live" hence the lack of an "interval between the production, transmission and reception of programmes" (Scannell, 1996, p. 105). In other words, the producers used it as a liveness indicator to create a simulated "live" that would replace the missing ontological liveness of the program itself.

Paratexts such as promos and commercial spots can also serve as strong liveness indicators since they belong to the live televisual transmission and coincide with the current airing of a program. Thus, they can similarly be used to infuse liveness into the programs they promote, even if they are not actually broadcast live.

The *Survivor* promos, for example, use present or future tense in their narration, even though the program is shot and edited many months before it is aired. Accordingly, a promo in the second season made reference to one of the most popular participants, addressing the viewers directly:

> Do you like Bashevkin? Can you do without him? They want to take him from you. Can you still save him? (Oct. 29, 2009)

Using the present tense, the promo implies that the show's events are still unfolding and the viewers can somehow affect the outcome. By drawing a symbolic temporal link between the televised (taped) events and their transmission, the producers constructed a "seemingly live" viewing experience which hints at one of the purposes of live TV: engaging the viewers with the text and highlighting their presumed impact on the show through para-interacive paths. This issue is explored at further length below. In the present context, however, suffice it to say that in addition to its obvious aim of promoting the show, the promo serves as a liveness indicator for the coming episode, and the series as a whole.

Similarly, an ad for a popular brand of Israeli breakfast cereal (Telma), which was broadcast repeatedly during the commercial breaks in the same season of *Survivor*, also served as a liveness indicator. Stating: "Telma wishes the survivors the power to win," it implied that the winner was yet to be determined at the time of the broadcast.

As noted above, *Romance Bay*, a locally adapted version of the American format *Paradise Hotel*, aired on Israeli TV in 2008 (Channel 2). A romantic reality show depicting beautiful young men and women seeking love at a photogenic holiday resort, the episodes were taped in advance, with several short live segments later added. As part of their effort to create the appearance of liveness and immediacy, the producers also shot an hour long behind-the-scenes program that was broadcast two hours before the first episode was aired (Sept. 3, 2008). It purported to be coverage of the program's set only hours before the live broadcast.

> Good evening and welcome to the hottest, sexiest set on the Mediterranean coast, Romance Bay. Right behind me you can see the final preparations for the sizzling new season of Romance Bay. . . . We don't have much time because the program will air in about an hour, so let's begin.

During another segment of this behind-the-scenes program, the host interviewed the producer of the series, asking:

> In an hour you are on the air [glancing at his watch] oh, actually a little less than an hour. . . . Isn't your heart racing?

The conversation drew a temporal line between the production process and the transmission time, implying synchronization between the two, even though neither the so-called "behind-the-scenes" edition nor the program itself were happening live. Pararexts such as behind the scenes programs, mold the viewing context of the whole series (Gray, 2010) while symbolically draw a contract between producers and viewers. Thus, even a moment before a show is aired, such programs can be used as liveness indicators when no real liveness exists.

Televisual "Flow" as a Liveness Indicator

Televisual flow was conceptualized as one of the main characteristics of broadcast TV (Williams, 1974). It refers to the way one text can be affected by, and even assimilated into, the following text, impacting the content and the viewing experience (Wood, 2007), with the aim of holding the audience from one program to another or one episode to another. Extending Jonathan Gray's (2010) concept of paratext I propose to regard the "flow" through which a certain program is transmitted as a paratext that gives meaning and context to program and its viewing experience. Thus I suggest that every program can be affected by the previous program, so that if one program is transmitted, or perceived to be transmitted live, such as a newscast for example, it can serve as a liveness indicator for the consecutive taped program. This pretend temporal link is often highlighted intentionally.

To demonstrate, the satirical shows *Eretz Nehederet* (Channel 2) and *The Back of the Nation* (Channel 10) are prerecorded shows that are aired immediately after the main evening newscast. At the end of the news, the newscaster generally turns to a large screen in the studio and greets the host of the satirical show, who, in turn, replies supposedly "live" on screen, even though this segment, like the program itself, is taped. The on-screen link between the live newscast and the following taped show infuses the latter with the conspicuous liveness of the former. The spurious temporal link between the two programs echoes the traditional role of television announcers, whose very presence and scripted words stressed the here and now of the television viewing experience (Van den Bulck and Enli, 2014).

A similar liveness indicator within the televisual flow was employed by Channel 10 for the first few seasons of *Survivor*. This show, too, aired immediately after the main evening news. At the close of the news, the newscaster turned back to a screen in the studio and announced: "And now for *Survivor*." By means of this simple gesture, the newscaster was able to add a layer of liveness to the taped program. In some cases, the newscasters appear to be preparing to watch the show themselves, mirroring the viewers at home. Moreover, the projection of the taped show on the screen in the news studio usurped the immediacy and liveness of the news items that had been shown on the same screen in exactly the same way only minutes before, tying the non-live program to the live news not only temporally, but spatially as well.

Graphics as Liveness Indicators

Graphic additions to the TV screen made their debut in the 1980s, when American television embraced the "televisual style," described by Caldwell (2004a) as video-graphic overload crowded together in the televisual frame. Very often, current televisual graphics are not an integral part of the frame, but rather attached to it to form what seems to be an additional visual and informational layer. In some instances, these graphic additions resemble the digital banners characteristic of online aesthetics. Some are clearly used as liveness indicators, the most iconic being the caption "live" or "direct" at the edge of the frame during a live broadcast.

One creative graphic evolution of the "live" caption is the *Big Brother* "eye" graphic that appears in the upper left corner of the frame for the most of the program. In the first season of the Israeli *Big Brother* (2008), the "eye" icon appeared next to the "live" caption to symbolize the liveness of the format. In the following season, the "eye" graphic became animated ("live" is, in effect, the essence of the concept of animation), resembling a human eye staring directly at the viewer. This "straight in the eye" graphic icon might be considered as a digital illustration to the look of the televisual hosts in TV shows such as talk shows or newscast, meant to impart a sense of liveness and co-presence with the viewers (Horton and Wohl, 1956).

Another temporal graphic symbol often used during prime time is a digital stopwatch, which is usually added to the television frame a day or two before a special televisual event, such as the debut of a new series or the final episode of the season. This animated temporal indicator has been used in many popular shows in Israel, including *Big Brother* and the drama series *Hatufim* (later adapted into the American series *Homeland*), among others. It serves as a liveness indicator since it implies a symbolic temporal synchronization between the television world and the viewer's world, inviting viewers to synchronize their watches with the television schedule in anticipation of the live transmission, even if the show itself is taped. The strategy is reminiscent of the analogue clock that was screened between programs in the early days of Israeli television (1960–1970) it was mentioned at the forward of this book. Generally appearing thirty seconds before the start of a show or newscast, the clock would fill the frame, ticking loudly as its hands advanced, calling viewers to sit down in front of their TV set and wait anxiously for the live broadcast to begin.

In the Israeli singing contest format *Rising Star* (2013–), several graphic liveness indicators were employed to emphasize temporal synchronization with the viewers. In this format, during the live auditions episodes, viewers could vote for a contestant using a special application on their phone, and the accumulating votes were graphically shown on the screen in the form of a rising column in what appeared to be real time. This strategy goes a step further than other live contests, where the votes are counted behind the scenes and the results announced later in the program since the counting seemed to graphically develop on screen in the shape of a graph. Being able to seemingly watch the votes being counted while the voting was still going on linked the viewers even more closely to the televisual temporality, framing the whole show as happening in real time and therefore enabling them to take part and impact it. Thus, this liveness indicator, actually more of a "real time indicator," also demonstrates the new adopted interactive affordances of the medium and the format, arousing associations with online games where users can affect the virtual environment in real time.

Last but not least, graphics can serve as liveness indicators for linear broadcasting schedule. Typically in the form of graphic announcements or banners at the top or bottom of the televisual frame, they remind viewers of upcoming scheduled programs and often invite them to visit online sites and applications and join in real-time program-related activities. These graphic additions symbolize televisual liveness on two levels: they suggest a temporal synchronization between the viewers and the TV transmission, a link that can no longer be taken for granted; and they contextualize the specific program it announces as "happening now," even when it is taped, by directing viewers to related real-time multiplatform activity. As we will see in the following part, online digital platforms are employed extensively by television producers today to enhance a program's liveness or produce an appearance of liveness where none exists.

Online Platforms: "Second Shift Liveness" and "Alternative Liveness"

Digital platforms such as internet sites or social media, which accompany television programs, enable producers to create a richer viewing experience (Bennet, 2008), to monetize content beyond the broadcast (Jenkins, 2006), to compete more effectively in a crowded mediascape (Sørensen, 2015). More importantly online platforms, can amplify the liveness of programs and extend their temporal existence between and beyond the actual broadcasts (Kjus, 2009a; Ytreberg, 2009). The following examples demonstrate how online platforms are used to "keep the program live between broadcasts" by the extensive use of liveness indicators and the production of "second shift liveness" and "alternative liveness."

"Second Shift Liveness"

The concept of "second shift" in reference to broadcast TV was first applied to the American police drama *Homicide*, broadcast in the 1990s (NBC). In an effort to attract viewers, the producers created an online site where the fans could supposedly watch what was happening at the police station that featured in the series after the TV broadcast was over, offering them a peek into police work on "second shift." In line with this concept of "second shift aesthetics" (Caldwell, 2003), I propose the term "second shift liveness" which denotes a temporal synchronized link between broadcast TV and its viewers via online platforms after the program, that is, during a "second shift."

The first example is drawn from the Israeli version of *Dancing with the Stars*. During the 2009–2010 season, the producers invited viewers to go to the show's Internet site to "look into" the rehearsal rooms in real time during the week between broadcasts, promising: "Live broadcasts from the rehearsal rooms. Every day you can watch the stars practicing, dancing, and sweating. And all live!"[5]

The additional online platform enabled "second-shift liveness" during the program's "afterhours," allowing viewers to watch the performers "off-screen" throughout the week. The site afforded an un-edited, direct look into the rehearsal rooms through what appeared to be a CCTV camera, enhancing the sense of immediacy and authenticity of the content. Moreover, it implied a synchronous ongoing off-screen relationship between the participants and the viewers, who could watch them live in real time and presumably without mediation on a "second shift." In the same line, twenty four hours before the live broadcast of the finale episode of *Survivor* (2009 season), viewers could watch the construction of the elaborate stage on the website, following in real time the progress made by the builders and the work of the lighting and sound technicians in preparation for the finale.

Taking "second-shift liveness" even further, the producers of *Survivor* advertised that viewers could watch "the survivors 24 hours online." Here the program itself was not even live. Nonetheless, viewers could presumably see what was happening on the island in "real time" even though these were taped and edited segments uploaded online. Thus, various types of multiplatform "real time" activities were used to imbue taped and edited program with nonexistent liveness.

Other shows, such as *Rising Star* and *The Voice*, make extensive use of digital social media, such as Instagram and WhatsApp, to create an ongoing temporal connection between the program and its viewers. Between broadcasts, pictures of behind-the-scenes real-time activities and of the contestants rehearsing are posted on Instagram. The photos are joined by direct appeals to the viewers to support a certain contestant, requests for their predictions of the outcome of the competition, and updates on what has been happening throughout the week.

In addition, social media was employed in an innovative and experimental manner in the first season of *Rising Star* (2013). The instant messaging application WhatsApp was used to create a virtual group that included a number of viewers and a "producer in the know." Dubbed "The Good Guys," members of the group were ensured a continuous link to the producer. In effect, although constructed to look like a friendly informal WhatsApp group, it actually operated only as a one-way messaging site from the producers to the viewers where messages and photos from behind the scenes were occasionally posted. Due to technological and production constraints, the group was abandoned by the production before the season was over (personal communication with Sh. March 17, 2014).

Social media like Facebook and Twitter are frequently used by programs aired live to create ongoing synchronous communications between viewers and the production team across platforms (for more examples from the BBC, see: Sørensen, 2016b). But they are also used by taped programs. The popular long-running Israeli investigative journalism program *Uvda* (Fact), for instance, is not broadcast live, yet at the end of the program the host invites the viewers to a live chat on its Facebook page or online forum where one of the reporters or an interviewee in a given journalistic story will "be there live" after the broadcast. Thus, "second-shift liveness," that is, online real-time engagement with the program, can be created even when the program itself is not live.

"Alternative Liveness"

A "broadcast" feature that was introduced by Facebook in 2016 enables users to transmit anything they wish live online at anytime. This offered TV makers the opportunity to produce an "alternative live broadcast" before, during, and after a show, whether live or taped.

Consequently, the morning show on Israeli Channel 10 started "broadcasting" on Facebook in May, 2016. During that spring, the hosts began broadcasting live online even before the start of the show as they were preparing in the studio, and continued to do so during the TV broadcast by means of a regular smartphone. Followers of the program's Facebook page even got a push message informing them that the hosts were "live on Facebook now." The message declared that it was an "alternative broadcast" that allowed viewers to ask questions, post comments, or simply "like" the live broadcast. The odd shooting angles, unsteady frame, and unclear sound made the online broadcast appear truly "alternative." During commercial breaks, the hosts talked directly to the viewers through the phone lens, provided additional explanations and details from behind the scenes, and replied to viewer's questions and comments.

On the morning of May 25, 2016, for example, the online broadcast began with the host, Or Heller, spending a considerable amount to time trying to adjust his smartphone on a small stand, struggling to keep it from falling. The result was a long blurry shot of his palm—hardly a televisual image. During the commercial breaks, the hosts were seen straightening their hair and clothes, joking with the people in the studio, or simply bored and yawning, while the sound man adjusted the microphones and the props man replaced the cups of water for the next guests—all live online.

Such online streaming produces an "alternative live" potentially enhancing the live TV broadcast by creating a continuous synchronous connection between the individual viewer and the show, the production team, and other viewers, liking or commenting on the Facebook page. It therefore promotes the program by highlighting its liveness and immediacy and affording viewers the chance to connect with it more intimately and without mediation.

Online strategies take what Bourdon described as the "real life temporality" that can be created by media coverage of the actors in a television show between episodes (Bourdon, 2000, p. 550) even further. Second-shift liveness and alternative liveness not only extend temporal synchronization with the viewers beyond the TV broadcast, but they also create this temporal link with the production team as well as the performers.

According to Ellis (2002): "Broadcasting unrolls in the present: this is its novelty as an aesthetic form. . . . They [the characters] live life along with us, but in a parallel world" (p. 75). Second-shift liveness and alternative liveness aim to build a temporal bridge between the two parallel worlds by keeping the show "alive" independent of its airing on TV. It would seem that liveness on television can no longer be seen solely in terms of the temporal unity between the event and its transmission. It now has to be understood in the context of the digital interactive environment as part of the dynamics between devices and content producers (Sørensen, 2016a).

CONCLUSION

Broadcast television faces many challenges in the digital age, and most particularly changes in the ways people watch TV. Services and technologies such as VOD and DVRs, not to mention companies like Netflix or Amazon and the ability to consume media content online, enable the personalization of viewing in temporal terms. Consequently, "live" television transmission is supplemented with the consumption of TV content in what Manovich (2001) dubbed "digital real time," which allows the viewer to "call up" content on demand (p. 223).These changes feed broadcast channels' fears of diminishing linear viewing and the collapse of the commercial TV model.

In an effort to counter this decline, TV makers are now taking advantage of the popular appeal of "live" (Lunt, 2004, p. 330), using live broadcasts to create "must-see-now TV." By luring viewers to sit in front of the TV set and watch the linear transmission, broadcast TV hopes to be able to maintain the business model of the thirty-second commercial spot. To the same end, "liveness indicators" are inserted even into non-live, taped programs.

Another aspect of contemporary liveness relates to the multiplatform mediascape. Live programs and "liveness indicators" within taped shows produce content "which makes a claim upon us to engage with it" (Auslander, 2012, p. 7). It attracts more audiences to more screens and encourages second screen engagement and related online conversation. Online engagement also attracts attention to a program, or as the TV industry calls it, creates a buzz around the show, which, in turn, draws advertisers. Furthermore, many broadcasters collect data on their viewers online behavior and apparently, the "mechanics of Big data suggests that its bias towards real-time analytics" encourages the production of live broadcasts (Kelly, 2017, p. 16).

Last but not least, "live TV" in whatever form, can be contextualized as a para-interactive strategy. In other words, it is aimed at forging and strengthening para-interactive paths by which audiences can participate in and make an impact on television texts, as discussed in chapter 3.

THREE FINAL COMMENTS ON THE MEANING OF "LIVE" TV IN THE DIGITAL AGE:

Live TV and liveness-infused taped programs are part of television's struggle to preserve its social and temporal relevance in the digital mediascape.

Couldry claims that ever since liveness became a quality of digital media, television has had to argue for its continued social centrality in the face of competing media (2003, p. 97). This is happening at the very time that television is losing its own liveness as the synchronization between distribution and consumption of TV content is diminishing (Kjus, 2009a). I have sought here to demonstrate how television enhances its liveness and maintains its temporal relevance as a medium not necessarily by following the liveness of events, but rather by constructing liveness indicators even where there is no ontological liveness.

Live TV highlights not only its connection to the real world, but also the connection between viewers and the televisual world.

Live broadcasting, as a major feature of television, was perceived in the past as connecting viewers to places and national or global events (Dayan and Katz, 1994), and to realities relevant to them as part of society (Couldry, 2004). Furthermore, their co presence with the television image gave viewers the sense that they were "witnessing" real events as they unfolded (Ellis, 2002). In addition, live broadcasts were seen to contribute to the intimate connection the viewers feel toward television personas (Horton and Wohl, 1956), both performers and the production team, and even to create "families where none exist" (Feuer, 1983, p. 20), as well as to link viewers to each other as a result of simultaneous shared viewing (Ellis, 2002; Dayan and

Katz, 1994). These traditional meanings of live TV are still relevant. However, in the context of digital media and the multiplatform media environment, it now also invites viewers to feel part of the text, to participate and engage with it and with the production team, to report on what they see, and to discuss it with other viewers online during and after the broadcast.

Live TV echoes "real time liveness" (Bennet, 2008), which affords real time interactivity similar to that of a computerized system.

Televisual liveness today imitates the "real time" online environment that is responsive to the consumer. Contemporary televisual liveness emphasizes the potential viewer's choice, impact, and participation in the televisual environment through para-interactive paths.

NOTES

1. Interview with David Wertheim, May 11, 2016. http://b.walla.co.il/item/2960144 (accessed: May 16, 2016).
2. Based on analysis of television schedules published on the site of the Israeli Second Authority for Television and Radio. http://rashut2.org.il/broadcast_list_archive.asp.
3. Yoni Porat, Mako Entertainment News, Dec. 3, 2009. http://www.mako.co.il/entertainment-tv-media/tv/Article-bc2644fca5e4521004.htm&sCh=3d385dd2dd5d4110&pId=237936823 (accessed: Dec. 5, 2009).
4. Mako, Sept. 9, 2016. http://www.mako.co.il/tv-the-mission-amazonas/season1-articels/Article-e74b55221964751006.htm (accessed: Sept.27, 2016).
5. http://reshet.ynet.co.il/11656.aspx (accessed: Dec. 2, 2009).

Chapter Five

Unveiling TV's Apparatus as a Para-Interactive Spatial Strategy

The theme of the live broadcast of the Emmy Awards ceremony in 2009, hosted by Neil Patrick Harris, was television's challenges in the digital age. The lyrics of the opening song outlined some of the industry's fears, especially the concern that viewers would desert the medium:

> Put down the remote, every note from his throat, is like a not-to-be-Tivo'd kiss.
> Don't jump online, 'cause this fine mug of mine, needs a huge high-def screen!

The threats to the TV industry referred to in the lyrics include delayed viewing, the option of skipping commercial breaks, and even worse, abandoning television altogether in favor of online content. The song framed the whole of the broadcast as a ceremony marking the achievements of American television in the context of these challenges. It took place on a huge, elaborate stage, and began with the host taking the viewers on a guided tour of its different sections:

> Harris [on stage]: I'll show you around. Big set, lots to look at. I'll tell you what's going on. [The conductor of the orchestra waves to the camera. The audience cheers.] Usually the band is in the pit so you can't see them. This year you can see them, so now you know why they're usually in the pit. [Harris points to a glass wall the band is sitting behind] Hey you all. Stop it, they can see you. No texting. . . . Over here we have the control room right here on stage. [Harris points to another glass room] Here it is . . . Glen Wise, our director there. Hi Glen! He's got control of all of these screens [close-up of the control technology] He's got all the shots. So pay attention! [1]

Here, the host describes the whole apparatus that was set on stage, and viewers are invited to watch it in operation. Thus, the band, which is usually an invisible source of music, can be seen on stage, and furthermore, the viewers are placed on a first-name basis with the director of the whole broadcast team members, who are generally unknown to them.

Traditionally, the control room is considered the "holy sanctum" of the TV production apparatus. Sealed off with heavy doors and soundproof, it is spatially separated from the stage and out of sight of the viewers at home. It is unquestionably considered a backstage area. When it was put on stage, the control room suddenly materialized as a "place," and more specifically as a place of work, with the viewers welcomed in and invited to observe what appeared to be the production process. The host's humorous warning to the production team, not to text during the broadcast drew a line between backstage behaviors and those appropriate on stage, while at the same time blurring the traditional border between these areas. This is just one example of a recent strategy whereby the production apparatus is made more visible to the home audience, and thus the line between back and front stage becomes less distinct.

I propose that although this is not an entirely new phenomenon, in its contemporary context in the digital age, it represents a para-interactive strategy increasingly employed by television producers. As I see it, this strategy is inspired by interactivity's promise of a potential role exchange between participants in a process of mutual discourse (Williams et al., 1988), and by the assumption of a participatory culture in a changed relationship between consumers and producers of media, one that is often based on reciprocity and equality (Jenkins, 2006).

The results of an extensive analysis I conducted of the programs aired on Israeli broadcast commercial channels (Channels 2 and 10) from 2008 to 2016 showed that most of the major prime time programs were accompanied by "behind-the-scenes" content. This included backstage footage, news items relating to the production process, and content manufactured by the production team that revealed the "backstage" area on social networks, especially Facebook. Moreover, some programs integrated traditionally backstage elements into the program itself, such as auditions, editorial meetings, or newsroom images. Although my research focused on Israeli television, the same trend can be found in other places around the world. One recent example is the internationally popular drama series *Outlander* (Sony, 2015–), which continually releases to its fans photos and short videos from the production set and information on the cast, including the casting process, and affords what seems to be direct contact with the writer of the series and the leading actors through numerous social media accounts. In addition, television's backstage and production process are the main theme of a number of fictional TV series, including sitcoms and dramas such as *30 Rock*, *The Newsroom*,

UnREAL, and more. In order to understand the various manifestations of this spatial para-interactive strategy, I suggest viewing it as a symbolic two-way route allowing for movement in opposite directions.

SPATIAL MODEL: SYMBOLIC BIDIRECTIONAL MOVEMENT

Unveiling television's apparatus may be perceived as a symbolic spatial strategy, and as such, it calls for a spatial model. A close look at prime-time television texts, including accompanying content such as promos (para-texts) and multiplatform content, reveals a situation that can well be depicted as a symbolic two-lane route: in one lane, television cameras move back and invite the viewers into the production space; in the opposite lane, elements that traditionally take place backstage, such as auditions, the artists' "green room," and routine activities in the newsroom, move forward to front stage, where they become an integral part of the performance. When combined, this symbolic bidirectional movement blurs the traditional spatial line between back and front stage and gives symbolic spatial representation to the interactive promise that media consumers can "trade places" with media producers. Moreover, the bidirectionality reflects a promise of symmetry and reciprocity between viewers and producers that is commonly associated with interactivity.

Backward Movement: Taking Cameras Backstage

Behind-the-scenes programs and segments, in which cameras enter backstage spaces, are very common on Israeli commercial television. They often accompany the opening episode of a season, showing parts of the production process and unveiling the apparatus: the control room, cameras, editing rooms, production team, etc. Additional behind-the-scenes footage can be found on videos on the channel's Internet site, promos showing preparations for an upcoming show, and more.

The following promo was broadcast on prime time just before the opening episode of the first season of *Big Brother* on Israeli TV:

> [Images of monitors in the control room, cameras, busy TV workers etc.]
> Voice-over:
> The house has been built. In the control room, which was only completed this week, the production crew is testing the 35 cameras that cover every corner of the house. Dozens of production team members are going through final rehearsals, getting ready to work around the clock.[2]

Here the cameras move backstage and the viewers are invited to enter the production space even before the broadcast begins. They can see for themselves the pre-airing preparations, the production equipment, and the way the things that make the show happen work behind the scenes. Similar to other supplementary content, promos are meaningful as para-texts since they shape the viewing experience of the programs they accompany (Gray, 2010). The *Big Brother* promo reports on the building of the set and the progress of the rehearsals, thereby aiming at involving the viewers in the production process and adding another perspective to their viewing experience, while at the same time strengthening the professional image of the TV crew.

Another example of the unveiling of backstage space can be seen in a segment that was uploaded to the Israeli News Corporation (Channel 2) Internet site:

> [A rapid montage of shots showing the newsroom in action]
> Voice-over:
> We wanted to show you behind the scenes of a newscast. Many viewers at home, watching the news at night, don't know how many people, news editors, reporters and producers, work on it during the day. [3]

The producers of the popular newscast invite the viewers into the newsroom, acquaint them with the "many people" that work on the news every day, and seemingly involve them in the professional behind-the-scenes efforts. As these two examples show, revealing the professional apparatus of television production aims not only increases the TV audience's engagement, but also emphasizes the skills and professionalism of the people behind the scenes.

Digital platforms constitute another avenue for inviting the viewers backstage. As demonstrated by the following examples, producers can take advantage of the synchronicity and immediacy of the online platforms that accompany television broadcasts to present viewers with live feed from the rehearsal rooms of popular prime time programs.

As mentioned in the previous chapter, during the 2010 season of the local version of *Strictly Come Dancing*, an online site offered a peek into the rehearsal rooms in between the live weekly broadcasts. Viewers could see what appeared to be live transmission from the cameras installed in the rooms. Since the transmission was continuous and unedited, there were times when all that could be seen were literally empty rooms. This, and the poor quality of the images and sound, created the impression of unmediated entry into the backstage area. Viewers were therefore afforded an "entry" into a typically inaccessible area, where hidden cameras enabled them to watch participants in sweat pants and no makeup displaying "behind-the-scenes" behaviors, such as squabbles, gossip, or just sitting around doing nothing.

In the Israeli version of *Pop Idol* (2010 season), the producers used a similar strategy, holding rehearsals in the original *Big Brother* house which was, naturally, fully equipped with cameras and microphones. Viewers could follow the rehearsals live, via the program's Internet site, whenever they chose. In both of these cases, viewers were invited to feel that they were continuously included in the production process, even between broadcasts.

The use of a digital platform alongside a traditional TV broadcast can also be illustrated by the now somewhat technologically outdated invitation issued to viewers at home to bring their laptops to the living room while watching the Israeli version of *Pop Idol* (2010 season). The host repeatedly encouraged the use of a special online site to "see the contestants behind the scenes" between the onstage performances, saying:

> Listen up! We have another innovation for you. This season the contestants won't rest during the commercial breaks. Come in and watch everything that's happening behind the scenes!
> [Commercial break]

Thus, viewers were invited to be in two places at once during the broadcast itself, both front and back stage, again blurring the boundaries between the two. Although critics might claim that the broadcasters were exploiting the performers even during their down time behind the scenes, turning it into a performance in its own right.

A more recent example employs Facebook's new feature of "live broadcasting." As described in detail in the previous chapter, the Channel 10 morning show has started transmitting what they call an "alternative" live broadcast on Facebook before and during its broadcast on television. One of the hosts shoots the preparations before the broadcast begins and then through the whole program with what appears to be his own smartphone, thus providing nontraditional shooting angles. This transmission is characterized by a narrow angle and an unsteady frame and the sound is not clear. During the commercial breaks the hosts talk to the viewers directly via the smartphone screen, describing the production process and adding details from behind the scenes. At other times during the show, the hosts were live on Facebook even during commercial breaks, when they were seen in typical "backstage" activities such as straightening their hair and clothes or simply yawning and looking bored. Viewers could watch other members of the production team, usually out of sight, working: adjusting microphones, replacing cups of water for interviewees or busy at the control room. During remote studio interviews, all the Facebook viewers could actually see was the host nodding or signaling to the floor manager, since the remote interviewee himself could not be seen or heard. Thus, the live streaming on Facebook offers viewers both an unmediated look at the production process and access to backstage, not only spatially but temporally as well: before the show begins and during commercial breaks.

The ambivalence towards the cameras moving backstage and inviting viewers to enter the production space and process surfaced in relation to the Israeli drama series, *Hatufim* (2010–2012), which gained international fame when it was adapted into the Emmy Award winning American series, *Homeland*. The producers created a video filmed on site that showed fragments of the production process and featured interviews with the creator of the series and the lead actors.[4] Viewers were invited to "join" the producers backstage and tour the production space. At the end of the video, however, one of the lead actors looks straight into the camera and says: "You don't really want me to tell you what's going on here." On one level, the actor is talking about not divulging the story line, but his comment is ambiguous and could also refer to unveiling the production process, which might shatter the illusion. This raises the issue of the television industry's tendency for confidentiality and concealment, a point I will return to later in this chapter.

The examples above illustrate the lane along which TV cameras move backstage, offering the viewers a look at places to which they are traditionally denied access. Taking the other lane in this bidirectional model, backstage elements are brought forward to the front stage area, to become an integral part of the performance.

Forward Movement: Bringing Backstage Elements on Stage

We are all familiar with the glass wall behind newscasters that brings traditionally backstage elements on stage, presenting us with a symbolic image of a busy newsroom. Indeed, in March, 2015, Israel's public channel (IBA) inaugurated an elaborate revamped news studio which boasted an animated image of a "newsroom" behind the presenters. During a tour behind the scenes with my students in January, 2016, they were surprised, and somewhat disappointed, to discover that it was not the actual newsroom, but merely a projected image of it. As one student remarked, when you watch the news at home, the background looks real and "makes you feel that you are part of the action."

Similarly, in 2011, Israel's Channel 10 was required by the regulator to relocate its news studio to the capital of Jerusalem. For financial reasons, only a small news studio was set up in the city, while the editorial staff and newsroom remained 65 kilometers away. In order to eliminate any sign of this distance, backstage activity in the newsroom was taped in its actual location and then screened behind the newscaster in the new Jerusalem studio during the live broadcast.[5] Viewers at home were supposed to believe that they were seeing not only the newscasters, but also the process of preparing the newscast behind them.

Going even further, during the 2011 season, Channel 10's Saturday night news magazine opened each episode with a simulation of an editorial meeting conducted in the studio in front of the cameras. Viewers could watch as the reporters and the host of the magazine discuss the evening's upcoming top stories. This (pretend) editorial meeting, which traditionally supposed to take place backstage some time before the live broadcast, was thus brought forward to become part of the show, both spatially and temporally.

In addition, numerous reality shows begin with a lengthy audition process presented on screen. Television formats such as *Pop Idol*, *Master Chef*, *The Voice*, and so on take the auditions, which were traditionally held behind the scenes, and bring them forward for all to see, knowing that auditions bring high ratings. The 2009 season of the Israeli version of *Pop Idol*, for example, included 12 full audition episodes, each 90 minutes long, and the 2015–2016 season of *Rising Star* began with 9 full episodes of auditions, all of which were broadcast during prime time. Lasting between 45 to 90 minutes, these broadcasts showed not only the contestants' skill or lack of it, but also their emotions and heart wrenching personal stories, unexpected moments such as a participant fainting,[6] and the judges "at work." As can be seen in the following dialogue conducted during a very bad audition in an episode of *Pop Idol*, the viewers were invited to observe a candid professional and personal discussion between the judges that revealed their personal relations and behaviors once considered suitable only for the backstage area:

Judge A: Hopeless.

Contestant [singing and mumbling]: I can't remember the lyrics.

Judge A: At least the tune is still . . .

[The judges try hard to not to laugh, but Judge A succumbs, bursting into loud and prolonged laughter, apparently unable to control himself]

Judge B: What's the matter with you?

Judge A: I'm sorry. I can't help it . . .

[Addressing the embarrassed contestant]: You killed me my friend . . . I apologize you are a great person.

Judge B [pointing to the contestant]: What can we do with him?

Judge A: Honestly, you cheered us up. We've been through hours and hours of auditions.

Contestant: I'm happy to hear that.

Judge A: You can win a prize as the number 1 out-of-tune singer.

Contestant: What can I do my friend?

Judge A: Quit singing! Thank you and good-bye.

[Voice of Judge A over images of the tired judges rising from their seats and leaving the stage]: After an audition like that, we desperately need a break and a breath of fresh air.[7]

Here we see several manifestations of backstage elements shown on stage. First, we are allowed to view the performance of an utterly incompetent contestant who has no chance of passing the audition, and who therefore will not take part in the actual on-stage competition. Secondly, the judges' harsh and humiliating comments, which might be expected to be best left behind the scenes, are offered for the viewers' amusement. And last but not least, viewers are afforded a look at the panel members appearing bored, frustrated, or enthusiastic, laughing hysterically, surviving an attack of the hiccups, and squabbling with one another, all "backstage regressive behaviors" that are not traditionally revealed to the audience (Goffman, 1959). This may explain why viewers seem to love audition episodes, whose ratings are often as high as the competitive performance episodes that follow.[8]

Even when auditions are conducted behind the scenes, videos of them are frequently released to the viewers when the program is already on-air. During the 2015–2016 season of the Israeli version of *Survivor*, for example, the production company released the audition videos of the contestants who made it onto the show. Very different in content and quality from the show itself, they included embarrassing moments and resembled amateur videos of the participants sitting on plain plastic chairs in what seemed to be the production offices, dressed in their everyday clothes and with no makeup.[9]

Along similar lines, the Israeli version of *Strictly Come Dancing* integrated scenes and interviews from the green room into the television show. The green room is meant to be a backstage space where performers can relax until they are required on stage. However, during these broadcasts, images of the performers lounging in the green room and interviews conducted with them there were brought forward and became an integral part of the show. Watching the performers' off-stage behavior during breaks seems to be an attractive addition to a TV show, and such segments are increasingly being broadcast, although the strategy is not entirely new. The producers of the Eurovision song contest, for instance, have been using it for the past few years. It has also been expanded, so that in the Israeli singing competition format *Rising Star* (2013–), backstage interaction is part and parcel of the format. Throughout the show, the two hosts stand in the wings of the stage and talk, seemingly between themselves, in an informal and supposedly spontaneous manner. From this backstage position, they criticize, praise, make fun of, and gossip about the contestants and the judges. From time to time they speak directly to the viewers at home, asking for their opinions, and thereby creating the impression that the viewers are included in these behind-the-scenes informal conversations.

We have seen, therefore, the bidirectional movement that takes the cameras back to make the production apparatus more visible on screen and brings behind-the-scenes action forward to make it an integral part of the broadcast. This spatial strategy blurs the line not only between back and front stage, but also between the producers and the audience, thus symbolically marking the cultural changes that enable ordinary people to become producers of media content.

BLURRING THE LINE BETWEEN VIEWERS AND TV PROFESSIONALS

While behind-the-scenes content is becoming ubiquitous, it is not a new phenomenon. Since its inception, the television industry has enabled the viewers an occasional peek behind the scenes, mainly in order to emphasize its professional expertise (Caldwell, 2008). In some instances "professional TV talk" not only highlights the abilities of the professionals, but addresses the viewers as "experts" as well, including them in the inner circle. In a behind-the-scenes program just before the start of the 2008 season of Israeli *Big Brother*, an interview with a person introduced as the "interactive department manager" contained so many professional expressions that it could have been a closed meeting in the production offices. The rhetoric included terms such as "formats" and "massive accompanying digital content," and noted how the online site "jumped" in other countries during the broadcast as did revenues, etc.[10] The segment therefore framed viewers as television savvy, as people who understand the production process and industry's interests and are familiar with the professional jargon. According to Gitlin (1993), audiences enjoy being treated as "know hows" and he critically explains this pleasure to false feelings of power and participation.

Behind-the-scenes programs might be compared to the "DVD extras" of the past, which blurred the lines between texts intended for PR purposes and those meant for the viewers (Brookey and Westerfelhaus, 2002). However, while DVD extras were generally aimed at film buffs (Skopal, 2007), such programs on popular broadcast channels invite wide audiences into the professional sphere of TV production.

Just as they were in the past, behind-the-scenes media interviews and videos are part of the public relations campaign designed to promote a program, as some of the producers I have interviewed openly admitted. Nevertheless, these same people contend that this is not the only, or even the main, reason for the popularity of these segments. Television professionals believe, and rightly so, that audiences find the opportunity to enter the backstage area extremely appealing, especially in the changing cultural environment.

THE VIEWERS' DESIRE TO BE INCLUDED IN THE PRODUCTION PROCESS

Previous research has shown that producers strongly believe that their viewers expect to engage with media texts (Enli, 2009; Sundent and Ytreberg, 2009). This was confirmed by the interviewees in my study (see chapter 2 of this book), who claimed that viewers not only seek engagement with media texts, but also want to be included in the very production process itself. In the words of a top executive at one of Israel's commercial broadcast channels:

> The shows with high ratings are those which the viewer feels he is a part of, part of the celebration, part of the production, part of what is going on behind the scenes, part of producing television maybe. . . . People are curious about how you make television. They aren't naïve anymore. (Interview with K., September 9, 2009)

Similarly, the editors of a satiric television talk show acknowledged in a newspaper interview that they aim to expose the show's production apparatus in response to viewers' desire to be part of it. As they explained:

> You expose the show's scaffolding . . . everything is out in the open. . . . You don't underrate the person sitting at home. . . . On one of the shows we pulled the teleprompter out in front of the camera. . . . The viewers can feel that they are not being cheated and that they can be part of what's on television.[11]

As this comment suggests, unveiling television's apparatus on screen is viewed as a strategy for producing a seemingly inclusive and reciprocal relationship with the viewers, where "everything is out in the open . . . and the viewers . . . are not being cheated." Producers today perceive television viewers both to be media savvy and to expect to be involved in, or at least informed about, the production process, and they act accordingly (Duffy, Liss-Mariño, and Sender, 2011).

The viewers' desire to feel part of the production process is particularly interesting in light of John Ellis's (2004) succinct claim: "[T]hose who receive television have little or no possibility of becoming producers of television utterances" (p. 276). Nevertheless, it is clear that in the current context of profound technological and cultural change, television professionals believe that viewers' expectations are changing and that they must adapt to the new reality. As we have seen, they are doing so by employing the spatial para-interactive strategy of unveiling the production apparatus, thereby inviting the viewers backstage, and allowing them to feel as if they, too, are part of the production process.

IS TELEVISION GOING THROUGH A "SPATIAL TURN"?

Goffman (1959) argued that backstage spaces are usually out of reach and out of sight, and therefore inaccessible to "outside people." Using Goffman as a point of departure, Meyrowitz (1985) focused on television's power to blur spatial, social, and cultural boundaries. His acclaimed book, *No Sense of Place* (1985), has been seen retrospectively acknowledged as instigating the "spatial turn" in media studies, or the call for a combination of media and spatial theories (Ericson and Riegert, 2010). This idea is worth revisiting at a time when television as a medium and a technology appears to be undergoing a spatial turn of its own on various levels. Television texts today migrate from one screen to another and one platform to another (Lotz, 2007). Media content is subject to a "space shift" (Jensen, 2006), no longer confined to a "box" in the living room. Along similar lines, it has recently been said that TV is being "relocated" (Gripsrud, 2010, p. 3) or converged (Caldwell, 2004a). My discussion of spatiality focuses on a different aspect: the spatial arrangements of broadcasting television's inner spaces and their visibility to the viewers. Following Scannell's (1996) observations regarding the production space and the TV studio as actual places, I propose that television now allows viewers access to several places at once.

First, as explained above, the spatial strategy that television has adopted emphasizes the spatial properties of the production apparatus and invites viewers to symbolically step inside the production space. Secondly, extending Meyrowitz's claim of television's power to blur spatial and social lines, I claim that television is now blurring its own internal spatial lines, particularly those between back and front stage and between producers and viewers, by offering viewers mediated access to places usually inaccessible to them. In other words, TV is allowing the audience to enter into its private place: the production area.

The question is whether television is changing its own spatial self-perceptions in response to changes in the wider media environment. TV is frequently portrayed as a "window to the world" (White, 2006), whereas digital and computerized media are often thought of as an "environment," or even a world of their own (Manovich, 2001). While the window metaphor implies an observer or onlooker, the digital world metaphors have spatial connotations, implying the potential for entry into or immersion in mediated virtual spaces (Ryan, 1994). I suggest that by employing a spatial para-interactive strategy, television presents its own production apparatus as a mediated space and invites viewers to enter and immerse themselves in that space, and in the production process in general. Therefore, we might say that television is not only a window to the world, but now also offers its audience symbolic entrance into its own private world.

THE "SPATIAL TURN" MEETS THE "DEMOTIC TURN": USING A PARA-INTERACTIVE SPATIAL STRATEGY TO CONSTRUCT INCLUSIVITY AND RECIPROCITY

Turner (2010) coined the term "demotic turn" to describe the increasing visibility of "ordinary people" on screen. As we have seen, producers believe that contemporary television viewers expect not only to be seen on screen, but also to feel as if they are taking part in the production of the televisual text. I contend here that the spatial para-interactive strategy described in this chapter is employed by TV producers in order to construct a viewing experience of inclusivity in the production process by simulating reciprocity between producers and viewers.

MacCannel (1973) maintains that people permitted entry into backstage areas (in the Goffmanian sense) of groups, cultures, and social institutions are, in effect, invited to become one of the group. By the same token, being permitted entry into television's backstage, the production space, implies inclusivity, with viewers invited to feel part of the production team. Furthermore, drawing on Goffman (1974), it might be said that by making the television apparatus visible and revealing the backstage on screen, TV producers are bringing viewers closer to the status of "participant" rather than "onlooker," thus cultivating their sense of inclusivity and involvement and implying the potential role exchange associated with interactivity (Williams et al., 1988).

In addition, as mentioned in chapter 3, the rhetoric of contemporary television no longer frames viewers merely as fans and media consumers, but also often allows them a degree of access into television texts as participants, whether as "narrative co-editors" when it enables them to determine the storyline through voting, or as "assistant writers" through a program's website (Andrejevic, 2006; Johnson, 2007; Ross, 2008). Moreover, many television texts are based on television cameras entering the viewers' lives. TV producers expect "ordinary people" (Turner, 2010) to allow them access to their homes, their kitchens, their families, and their social and romantic lives—in other words, to their own "backstage" (Evans, 2005).

I suggest this strategy is meant to further an appearance of reciprocity aimed at persuading the viewers of a more "equal" relationship between them and the producers. Reciprocity, which means that people have to "give what they want to receive" (Cialdini, 2001), is a basic feature of human behavior that comes into play in the process of persuasion. Therefore, while ordinary people share their backstage with television producers (and the viewing audience), television producers share the backstage of their professional world with the viewers, thus creating a sense of reciprocity that might imply greater equality between the two sides.

To summarize, unveiling television apparatus on screen is a para-interactive spatial strategy used by the TV industry as a simulation of a changed relationship with its audience. The simulated relationship is one of reciprocity, and includes the viewers in the production process in a manner meant to parallel the "interactive promise" of digital media.

IS IT REALLY BACKSTAGE?

Paradoxically, as television's backstage becomes more visible on screen, the industry is increasingly hiding behind real and legal walls of confidentiality. TV performers and employees are required to sign confidentiality agreements, and even members of the public auditioning for shows, as well as visitors behind the scenes, have to sign nondisclosure agreements before being allowed into the production area. This concealment is fueled by fierce commercial competition, along with restrictions imposed by global formats. The contemporary television industry thus not only produces illusions, but also invests a great deal of resources in constructing the illusion of reality, which often requires concealing production processes.

As only to be expected, walls of confidentiality occasion gossip and scandals. In 2010, an anonymous Israeli blogger posted embarrassing details from behind the scenes of popular prime time programs. An entertainment journalist covering the incident called the blog "a source of information and details that all fall under the category of what the people in the television industry don't want you to know" (Turovitz, 2010). The revelations seriously unsettled the industry, with some employees even being forced to take polygraph tests. It eventually turned out that the blogger had obtained his information by hacking into the production company computers and when discovered, he was arrested (Greenzweig, 2011). Such exposés of the "behind the scenes of behind the scenes" raise nagging questions about the true nature of the backstage that TV chooses to reveal to the audience.

Control of inside information is essential in order to guard the commercial interests of broadcasting channels and production companies. However, it would be impossible to conceal information while at the same time revealing carefully selected details that serve the broadcaster's interests without the cooperation of the entertainment journalists covering the industry. Although this cooperation is rarely made known to the viewer, the subversive and illegal activity of the blogger/hacker above prompted an industry journalist to publicly admit the existence of this "give and take" relationship in her column, stating:

> I suggest we don't pretend to be naïve. We need them and they need us . . . we take advantage of their popularity and in return we don't strip them bare in public. At least we don't expose those details which might affect ratings. (Turovitz, 2010, p. 15)

While most journalists cooperate with the PR mechanisms surrounding the programs, occasionally the audience is granted a sneak peek into the "behind the scenes of behind the scenes." The following article, describing the auditions for the popular Israeli version of *Pop Idol*, known as *A Star is Born*, is one example:

> He Did Not Pass: Pseudo Reality in A Star is Born
> Yoav Feldesh escorted his friend to the audition of the show A Star is Born and discovered how the efficient, brutal, and secretive system operates. Here the only reality is the one the production fabricates, and power rests with the camera. . . . It's a machine . . . and this column is a report from within. . . . Slowly it becomes clear that the role of the production assistant is to arrange the world for the camera. She knows what the camera wants. From the sidelines, it seems to be a bit infantile: she tries to dramatize reality. She tells a religious girl to go into the audition. The girl stands up. "What? You don't want to pray a little before you go in? Just to make sure you succeed?" the researcher asks. The girl says no. "Maybe you can pray just a little?" she pleads. "We're on TV, we have to see things." The girl refuses again." (Feldesh, Y. June 6, 2010)[12]

As explained above, auditions, which traditionally belong backstage both spatially and temporally, are now made part of the show to give the viewers the sense that they are observing the spontaneous and unscripted activities behind the scenes. It is obvious from the description above, however, that the activity revealed to the audience is actually carefully controlled, staged, and scripted by the producers. Those who have come to audition are expected to play their part, as dictated by the interests of the producers and the demands of the narrative, even before they go on stage. In effect, they are directed and staged as soon as they enter the "backstage" area, long before the show starts and even before they have passed the audition and become full-fledged contestants.

"STAGED BACKSTAGE"

MacCannell (1973) speaks of tourists' eternal search for authentic experiences and original sites in a new place or culture, describing their desire to visit "back" areas as an important part of the "authentic experience" they seek. Entering "back regions" means according to MacCannell is "Being 'one of them' or at one with them . . . this is a sharing which allows one to see behind the others' mere performances" (p. 592). However, one of his main arguments is that tourists are often guided to places that are actually designed and staged to look like backstage, such as the "authentic" village, the "typical" kitchen, or the "traditional" ceremony staged especially for visitors. Similarly, I claim that inviting viewers to enter television's backstage is only a *para*-interactive strategy. In fact, similar to tourists, viewers are led through nothing more than a "staged backstage." It is a staged experience that is intended to construct an illusion of inclusion and reciprocity tailored for the viewers at home. In effect, it is part of the performance, carefully constructed according to production interests. While the apparatus of television is seemingly becoming increasingly visible on screen, the actual backstage is protected behind thick walls.

CONCLUSION

Goffman (1974) defines two different "information states" that characterize people in the context of a theatrical play. The actors and production people are in one information state, while the audience is in a different information state, since they have less information about the unfolding play. I maintain that by ostensibly unveiling the televisual apparatus to the viewers, producers are trying to give the impression that the information state of the audience is closer to that of the production team. Aware of the viewers' desire to feel part of the production process, they invite them into what appears to be behind the scenes in order to create a sense of inclusion and equality. In Goffman's terms, once the viewers are supposedly in the same information state as the production team, that is, behind the scenes, they become less "onlookers" and more "players," in this case participants in the production process rather than actors on stage. Narrowing the gap between the different information states hints at a potential role exchange between producers and viewers, and echoes one of the features of interactivity, the possibility of an exchange of roles in the interaction process (Williams et al., 1988). While exposing what happens behind the scenes is not a new idea, in the changing media environment, where the lines between producers and consumers of media are blurred, unveiling television's apparatus on screen and inviting viewers to be part of the production process becomes charged with additional meanings.

To conclude, I would like to return to the same Emmy Awards ceremony with which I opened this chapter. Placing the control room on stage illustrates my main arguments and demonstrates the spatial para-interactive strategy discussed above. In this live broadcast, the control room was brought on stage (one could say that it was literally "staged") behind a glass wall, and the host invited the viewers to "supervise the production team's work." Thus, the visibility and accessibility of the production process were highlighted. However, here, as in the other examples I have presented, the audience was granted no more than a partial and fleeting glimpse of the actual production apparatus.

Spatial divisions and borders, whether physical or symbolic, construct social, cultural, and organizational hierarchies and power relations (Caldwell, 2004b; Couldry, 2001). In this chapter I have shown that although television employs the spatial para-interactive strategy of unveiling its production apparatus, permission to go behind the scenes is still strictly limited, and viewers are not really given actual access to the means of production. Producers create a simulated "backstage" on stage, keeping the actual behind-the-scenes area tightly secured and sealed off. This strategy, then, is designed to create an illusion, but does not, in fact, change viewers' traditional role as onlookers (Goffman, 1974). Unlike the role exchanges in the digital environment it is supposedly seeking to emulate, broadcast television continues to maintain the established power relations between producers and viewers.

EPILOGUE: THE TRANSPARENT STUDIO

One cool summer evening in June 2010,[13] I was strolling in the center of Jerusalem. It was around 8:00 p.m. and the streets weren't crowded. I walked through a small stone paved alley and reached Zion Square in the center of the city. As the square opened up in front of me, I saw a huge brightly lit glass cube surrounded by metal barricades and a few bored security guards. Coming closer, I realized that it was the well-publicized "Transparent Studio" of the music cable TV station, Channel 24. Since 2003, this Israeli channel, owned by Keshet which also operates the country's most popular commercial channel, broadcasts music clips, entertainment programs, and singing contests. The Transparent Studio was a project that ran for several years, most often during the summer. Each time the glass-walled studio was erected in a different location, such as the beach or the center of Tel Aviv. In the summer of 2010 it was broadcasting from the center of Jerusalem, transmitting live performances and light talk shows. Since the walls were transparent, the people outside could see not only the hosts, the singers and bands, and the talk show guests, but also the production apparatus at work. Thus, all the lighting, cables, electricity generator, cameras and camera men, audio equipment, monitors, etc. could be seen through the glass walls. That evening, I could see inside the glass structure, the recent winner of *Big Brother* Israel,[14] turned TV host, who was interviewing a local female celebrity.[15]

They were chatting pleasantly, but I had a hard time hearing what they were saying. The audio from the TV show was transmitted to the street through a small loudspeaker that emitted nearly unintelligible sounds.

As I looked around, I noticed that, except for the security guards, only a few people were standing near the studio watching the show. At first, I thought the reason for the tiny audience might be that it was the time most people were eating supper, or the unusually cool weather, or even the fact that Jerusalemites are known to be homebodies. And then the sound of loud music not far away attracted my attention.

Going over to investigate, I saw a large group of people cheering and clapping as two masters of Capoeira performed the Brazilian martial art. The enthusiastic crowd was clearly caught up in the music and the daring routines of the performers. I stood there for a while as more and more people came to watch, and then turned to look back at the almost deserted glass studio nearby. It was obvious which of the two attractions was more appealing to the passersby in the center of Jerusalem.

All the while the lively music echoed through the square, inside the transparent studio things went on according to the planned "line-up," and neither the host nor his guest seemed to be aware of the commotion outside. This isn't surprising, as TV studios are generally sealed to external noises, and the bright lights required for the broadcast would have turned the glass walls of the cube into a mirror for those inside. So it's no wonder they were oblivious to what was happening just beyond the studio walls.

Placing the transparent studio in the town square, that is, in the "viewers' world" rather than "TV's world" according to Couldry (2001), symbolizes television makers' aspiration to unveil the TV apparatus. The glass walls offered a view of both the program and its "behind the scenes." Meant to be a participatory experience, it aimed to blur the spatial line between viewers and producers. Yet, as I stood next to the glass cube that evening, I could not help thinking that it was a perfect example of the essential features of the televisual medium: one-directional, linear, unable to create genuine reciprocal interaction with its audience, and closely adhering to its predetermined programming. TV may attempt to appeal to its audience by appearing to be open and inviting, but as the transparent studio demonstrated, it is fundamentally incapable of blurring the line between the producers and the audience, due, in part, to the technical requirements of the medium. Thus, in many senses it is still a one-way medium that does not afford viewers' participation in the production process. Consequently, despite the efforts of television makers today, the traditional line between front and back stage and between media producers and consumers continues to exist.

NOTES

1. Transcript from http://www.buzzsugar.com/Video-Neil-Patrick-Harris-Singing-Dancing-Opening-2009-Primetime-Emmys-5145139 (accessed May 2, 2010).
2. *Big Brother* promo, broadcast September 1, 2008, Channel 2.
3. Online video, produced by the Channel 2 news company http://www.youtube.com/watch?v=Unoplod7RV8 (accessed November 15, 2011).
4. Online video, produced by Keshet, http://www.mako.co.il/tvhatufim/behind.html (accessed March 20, 2014). The video was never broadcast on television because of changes in programming due to the Passover holiday (private communication, March 27, 2014).
5. Private communication, March 13, 2012.
6. *Pop Idol*, Season 8, May 5, 2010.
7. *Pop Idol*, Season 8, May 6, 2010.
8. The audition episodes of *Pop Idol* in May, 2010 recorded an average 24 percent rating, http://www.midrug-tv.org.il/scripts/rating10.asp (accessed July 18, 2010).
9. http://reshet.tv/Shows/Survivor/videomarklist,243543 (accessed April 7, 2016).
10. Interview with Yuval Natan, Channel 2, September 1, 2008.
11. Izikovich, 2010.
12. Yoav Feldesh, *Calcalist*, June 6, 2010. http://www.calcalist.co.il/articles/0,7340,L-3403656,00.html (accessed July, 27, 2010). Additional examples of journalistic articles and independent blogs that partially expose the real behind the scenes of popular television shows include:
    http://www.mouse.co.il/CM.television_articles_item,789,209,50701,.aspx
    http://www.tapuz.co.il/blog/viewEntry.asp?EntryId=1428160
    http://www.mako.co.il/news-channel2/Friday-Newscast/Article-0a31ea617445d11004.htm
    http://www.haaretz.co.il/misc/1.1190824
    (all accessed April 21, 2016).
13. Tuesday, June 15, 2010, at 8:00 p.m.
14. Eliraz Sadeh, who won the 2010 season of *Big Brother* Israel.
15. The actress Ilana Berkowitz.

Conclusion

With the dramatic changes in the mediascape and the spirit of the "interactive promise" and "participatory culture" looming over television, producers know they have to "get out of the box" and adapt to the change in power relations between producers and consumers of media. I have sought here to identify and chart the changes television is going through and the ways it is trying to appeal to its audience at a time when viewing habits and expectations from TV are changing. But I did not set out on this exploratory mission empty handed. At my disposal were maps already charted by past researchers, such as Horton and Whol (1956) among others, which helped me to remember that the televisual strategies I identified in contemporary TV have their roots in the legacy of the medium, including para-social relations, "live TV," and "behind the scenes" texts. I discovered that the new meets the old on various dimensions, and many of the contemporary strategies employed by TV today were drawn from previous strategies and adapted to current conditions and challenges, thus giving new relevance to what came before.

The Israeli TV industry proved to be a fascinating terrain for this explorer. Relatively small, fiercely competitive, and under continuous commercial pressure, it can serve as an excellent case study as it is facing similar challenges to many other broadcasting channels around the world that are now operating in conditions of uncertainty.

This book tells the story of contemporary broadcast TV from two main perspectives: that of TV makers, and that of the texts they produce. Thus TV makers' doxa led me to study the praxis, their work (Thompson, 1990). I began by exploring the views and professional perceptions of TV makers and changing organizational norms through interviews and participatory observations. I discovered an industry in turmoil, operating under conditions of uncertainty and anxiety, with the technological, cultural, and commercial ground moving under its feet. The concept of "interactivity," a ubiquitous term in the popular and academic media discourse, was invoked in the discourse of TV professionals as well. It was often described as a significant characteristic of the "new digital" media environment and as one of the main features of its dramatic transformation. The lack of the potential for interaction with the viewers was perceived as one of TV's main weaknesses in the race against online applications and social media. Even though they were acutely aware of the competition they face, TV makers expressed duality and ambivalence toward the medium's relevance and the question of whether or not it has to change. They were also ambivalent in regard to their professional skills and the value of their cultural capital as creative professionals in a changing cultural atmosphere.

One thing about which television professionals are not ambivalent about is the understanding that viewers' expectations are changing. They are aware of changing media content consumption habits as a result of services such as VOD and DVRs, not to mention companies like Netflix or Amazon, and the ability to consume media content online, enabling the personalization of the content as well as the temporality of viewing. These changes feed broadcast TV's fears of diminishing linear viewing and the crumbling of TV's traditional commercial model. Interactivity and the resulting cultural phenomenon of "participatory culture" are perceived by TV makers to be significantly affecting TV viewers' preferences and habits. They believe that viewers today seek interaction with TV texts and potential involvement in the production process, similar to the affordances of the interactive communication environment in which they can easily become producers and distributors of media content, often from their own homes. Consequently, they typically voice the opinion that TV too should afford more interaction with its audience and allow some degree of involvement, or at least a peek into the production process. These findings were discussed in chapter 2.

These perceptions regarding the desirability of interaction and interactivity as a symbolic concept was the departure point for the second perspective explored in this book: the televisual texts of commercial broadcast prime-time television and accompanying online content. Combining TV makers' perceptions with a close inspection of prime-time programs enabled me to identify and conceptualize a cluster of para-interactive strategies aimed at appealing to viewers by presenting them with various invitations for actual, as well as feigned, interactions with TV and its texts and producers. Although para-interactive strategies are based on different elements of interactive communication processes, they are closely interrelated. They can be depicted as a typology of televisual strategies over three dimensions. The first are *para-interactive paths* that invite viewers to interact with television by responding to, participating with, or making an impact on the program or engaging with accompanying texts through various digital platforms. The para-interactive paths TV adopts are inspired by the interactive promise of participation and the impact media consumers can have on media messages. However, these paths offer limited participation and opportunities to impact the message only in well-defined and restricted manners. These paths were discussed in length in chapter 3.

The second dimension are *temporal para-interactive strategies* aimed at producing an ostensibly "live" or "real-time" viewing experience in order to create a temporal link between televisual texts and viewers. The purpose of the temporal strategy is to attract viewers to linear TV transmissions and strengthen the para-interactive paths by creating the impression that the program is presumably still unfolding and can therefore be shaped and impacted by the viewers. This dimension was discussed at length in chapter 4.

The final dimension are *spatial para-interactive strategies*, which unveil television's apparatus on screen in order to construct the appearance of viewers' involvement in the production process and hint at a relationship of inclusivity, reciprocity, and equality between the viewers and the producers. The production apparatus is unveiled through the bidirectionality of discursive and visual elements: in one direction, cameras (and audiences) are invited backstage, and in the other, "backstage elements" are brought forward to become an integral part of the program. Nevertheless, permission to go behind the scenes is strictly limited, and viewers are not really given access to the means of production. Instead, producers create a simulated "backstage" on stage, keeping the actual behind-the-scenes area tightly secured and sealed off. This strategy, then, is designed to create an illusion. Thus, while broadcast television purports to emulate the role exchanges between producers and consumers of media in the digital environment, it continues to maintain the established power relations between TV makers and viewers. This strategy is discussed in chapter 5.

All these para-interactive strategies contain features or elements that echo interactive communication processes and participatory culture. However, when employed by and adapted to TV, they do not generally construct actual interactive communication, and hence the term "**para**-interactive." TV makers contend that these strategies can potentially attract a larger audience, encourage second screen engagement, and generate related online conversation, which in turn creates a buzz around the show and draws advertisers. Whether or not they are right, it is important to note that even though para-interactive invitations are woven into or accompany many televisual texts, viewers are not obliged to respond to them. All programs can still be watched and enjoyed in the traditional "passive" way.

PARA-INTERACTIVITY AS A HYBRID CONCEPT

Para-interactivity may be viewed as a hybrid concept since it employs televisual, discursive, and visual elements known to TV for decades in order to produce the impression of "new media," that is, the interactive communication process typical of digital media. Metaphorically speaking, the concept has its head in the digital era and its feet in traditional television production conventions. As far as broadcast TV executives are concerned, para-interactivity is a winning formula because it allows them to feel innovative, contemporary, and competitive while still employing TV strategies that are part of their familiar professional tool kit.

An advertising campaign for a commercial broadcast channel and its new online site epitomized, both conceptually and visually, the coveted hybridization of TV in the digital age. The campaign promoted the channel as "Intervision," Internet and television combined, and its slogan was: "Enjoy all worlds."[1] The "worlds" it was referring to were the two technological/cultural worlds: the old world of television and the new world of the Internet. The campaign suggested that they were fused together, complementing and cooperating with each other. The visual images were imagined animal hybrids: half rhino and half rabbit; half peacock and half horse, etc. The combination of Internet and television was promoted as enabling viewers not only to access content at will, but also to interact with the texts and with other viewers.

Although this type of hybrid creation may be the goal, I would suggest that para-interactivity is a different sort of hybrid. It is a collection of televisual strategies born in the "television world," a world of mass media and linear top-down transmission of content run by people who "know how to make television" and at the same time believe that their viewers, especially the younger ones who live and breathe the digital world, expect "to take part." Para-interactivity was developed within the cultural and aesthetic world of television. It can be seen as the older medium's "training wheels" as it sets out on its bumpy ride toward the new interactive world.

In this context, para-interactivity affords TV certain advantages. First, it can help bridge the "participation gap" (Jenkins, 2006, p. 23) between people who are not comfortable with new technologies and those, mainly younger audiences, who participate and interact online. As noted above, para-interactive invitations do not have to be accepted for viewers' to feel as if they do, or at least can, participate. Moreover, accepting the invitation, by voting for your favorite contestant in a singing competition for example, does not demand technological literacy.

In addition, para-interactivity allows TV to place itself metaphorically between the older linear mass media and the newer interactive media. Thus television can seemingly be part of the transformation of the mediascape while at the same maintaining its professional and organizational status quo. In other words, para-interactive strategies enable TV to look as if it grants the viewers power and visibility without actually changing the power relations between producers and audience. Moreover, by employing these strategies, television today can invite viewers to engage in an individualistic interaction with the text while at the same time enabling them to participate in a shared TV viewing. The individual can choose whether to multitask on a second screen during the TV broadcast or to remain a "couch potato." Thus, through para-interactive paths, viewers can feel active and involved without investing too much effort, whether they accept the proffered invitation or not. And when they do, the extent and nature of their participation remains firmly in the hands of the producers.

CULTURAL AND SOCIAL IMPLICATIONS OF PARA-INTERACTIVITY

The implications of para-interactivity can be seen on three levels: financial, cultural, and social. First and foremost, para-interactive strategies are part of television's struggle to remain financially viable by preserving its relevance in the digital and interactive mediascape. With people now viewing TV differently, linear broadcasts are losing their audience, giving rise to the fear of lost revenue from advertising. Para-interactive strategies are therefore aimed at encouraging viewers to watch live transmissions and to "take part" in and create a "buzz" around them. Television is not the only traditional medium to employ strategies of this sort in an effort to shore up revenues. Many newspapers now also have online versions containing interactive elements. In some instances, this approach has proven successful. For example, newspapers in the United States that aggressively adopted hyperlinking practices were found to have a greater likelihood of survival in the long run (Weber and Monge, 2014). Within the affordances of "online environments," interactive practices such as hyperlinks, comments, likes, and shares can be easily combined. Television's use of online platforms with interactive affordances, however, can only come as an addition to the broadcast. Nevertheless, para-interactive strategies allow TV to imitate interactivity within a noninteractive, linear media environment. Can broadcast TV rely on para-interactivity to retain its viewers and ensure its survival? Or are these strategies only the first tentative steps on TV's journey toward a new digital world? It is still too early to answer these questions with any certainty. However, TV makers are convinced that the strategies they are now employing are effective in attracting viewers. That said, it must be noted that they are also adopting additional practices in order to preserve the relevance of broadcast TV. These include the production of high quality TV series with high "production value," the creation of online content aimed at attracting digital ads, and last but not least, the development of original and innovative TV formats intended for global distribution.

The cultural implications of para-interactivity relate to the fact that it is allied with contemporary developments such as "participatory culture," which demand redefinition of the relationship between traditional media and its audience (Jermyn and Holmes, 2006, p. 55). Even before considering the impact of para-interactivity in this context, we must first ask whether interactivity itself can, in fact, empower the audience and change the power relations between the consumers and the producers of media. The more optimistic opinion holds that, at least in some cases, the media can actually activate the audience and encourage its empowerment in a manner that might also be translated to the civil sphere (Coleman, 2006; Kim and Sawhney, 2002; Van Zoonan, 2004). On the other hand, there are those who perceive interactivity as merely a commercial transaction (Tincknell and Raghuram, 2002), increasing the profits of the producers while leaving power relations unchanged (Andrejevic, 2004a, p. 44). On the assumption, therefore, that interactivity does not necessarily entail a change in power structures, its derivative, para-interactivity, can not be expected to facilitate true empowerment of the viewers or bring about any real change in power relations between them and producers. As chapter 3 shows, para-interactive paths typically produce only the semblance of an invitation to the audience to take part in the program. Similarly, the supposed invitation to the viewers to take part in the production of the program through the "unveiling" of the production apparatus, discussed in chapter 5, is actually only the construction of a "staged backstage." Thus TV viewers can not really become producers, and para-interactivity grants them no more than "pseudo-power" (Holmes, 2008b, p. 15).

Finally, as described in the forward of this book, broadcast TV has always played a significant social role. It still connects people to places and national or global events (Dayan and Katz, 1994), and to realities relevant to them as part of society (Couldry, 2004). Furthermore, the television image can give the viewers the sense that they were "witnessing" real events as they unfold (Ellis, 2002). In the context of digital media and the multiplatform media environment, however, para-interactivity highlights the connection to the televisual world more than to the real world. Hence, these strategies might be said to frame TV not as "a window to the world," but as a window to itself. It now invites viewers to feel part of the program, to participate and engage with it and the production team, and to report on what they see on the screen and discuss it with other viewers online during and after the broadcast. Although this might be seen critically as isolating the viewers from the real world or the surrounding social reality, I contend that para-interactivity might indirectly be socially instrumental as well. By helping broadcast TV to retain its linear viewership, even if only temporarily, para-interactivity might preserve TV's role as a medium that affords shared viewing by the family, the community, and the nation. Looking back, it may be remembered that when broadcast TV was first losing viewers to multichannel cable TV, there were those who warned against the social consequences of fragmenting the public into smaller segmented audiences. The new digital technologies go even further, fragmenting the audience into individual content consumers who personalize their own viewing experience. From a normative perspective, broadcast TV's continued relevance has social significance as a source of information and an arena for public debate. Its survival, therefore, is not only a matter of financial self-preservation, but is also in the public interest.

FUTURE RESEARCH

Para-interactivity is demonstrated here mainly through the case study of contemporary Israeli broadcast TV. Analysis of TV in other places around the globe could reveal whether there are cultural variations in television's adaptation to contemporary technological and cultural changes.

Another question for future research is whether para-interactivity, as a set of strategies, is unique to television. It would be interesting to explore the strategies employed by other traditional media, such as radio or newspapers, to appeal to their audience and adapt to the digital environment.

Furthermore, I studied television through two complementary perspectives: that of its makers and that of its texts. Still missing, however, is deeper understanding of the viewers' expectations and perceptions of TV and its adaptation to the digital era.

Last but not least, TV, both as a medium and as an industry, is constantly evolving. My research project focused on a specific period in time (2008–2016). Continuing research efforts will be able to trace future developments in televisual texts and changes in TV makers' perceptions, as well as the introduction of additional innovative para-interactive strategies.

NOTE

1. http://flix.tapuz.co.il/v/watch-2739889-.html (accessed: Feb. 27, 2012).

References

Abramovitz, Yuval. (2010). "Eliraz Speaks." *Israel Hayom,* March 12.

Andrejevic, M. (2002). The kinder, gentler gaze of big brother: Reality TV in the era of digital capitalism. *New Media & Society, 4*(2), 251–270.

———. (2004a). *Reality TV: The work of being watched.* Lanham, MD: Rowman & Littlefield.

———. (2004b). The webcam subculture and the digital enclosure. In N. Couldry and A. McCarthy (Eds.), *MediaSpace: Place, scale, and culture in a media age* (pp. 193–208). New York: Routledge.

———. (2006). The discipline of watching: Detection, risk, and lateral surveillance. *Critical Studies in Media Communication, 23*(5), 391–407.

———. (2008). Watching television without pity: The productivity of online fans. *Television & New Media, 9*(1), 24–46.

Ariel, Y. and Avidar, R. (2015). Information, interactivity, and social media. *Atlantic Journal of Communication, 23*(1), 19–30.

Askwith, I. D. (2007). *Television 2.0: Reconceptualizing TV as an engagement medium* (MA thesis, Dept. of Comparative Media Studies, Massachusetts Institute of Technology, Cambridge, MA).

Auslander, P. (1999). *Liveness: Performance in a mediatized culture*. New York: Routledge.

———. (2008). *Liveness: Performance in a mediatized culture*. New York: Routledge.

———. (2012). Digital liveness: A historico-philosophical perspective. *Journal of Performance and Art, 34*(3), 3–11.

Auter, P. J. and Davis, D. M. (1991). When characters speak directly to viewers: Breaking the fourth wall in television. *Journalism Quarterly, 68*(1/2), 165–171.

Avidar, R. (2013). The responsiveness pyramid: Embedding responsiveness and interactivity into public relations theory. *Public Relations Review, 39*(5), 440–450.

Baudrillard, J. (1985). The masses: The implosion of the social in the media. *New Literary History, 16*(3), 577–589.

Bennett, J. (2008). "Your window-on-the-world," the emergence of red-button interactive television in the UK. *Convergence: The International Journal of Research into New Media Technologies, 14*(2), 161–182.

Bolin, G. (2009). Television textuality. *Nordicom Review, 30*(1), 37–53.

Bourdieu, P. (1996). *The rules of art: Genesis and structure of the literary field.* Stanford, CA: Stanford University Press.

Bourdon, J. (2000). Live television is still alive: On television as an unfulfilled promise. *Media, Culture & Society, 22*(5), 531–556.

Boyle, R. (2014). Television sport in the age of screens and content. *Television & New Media, 15*(8), 746–751.

Brookey, R. A. and Westerfelhaus, R. (2002). Hiding homoeroticism in plain view: The fight club DVD as digital closet. *Critical Studies in Media Communication, 19*(1), 21–43.

Bruns, A. (2006). Towards Produsage: Futures for user-led content production. In F. Sudweeks, Herbert Hrachovec, and Charles Ess (Eds.), *Proceedings: Cultural attitudes towards communication and Technology* (pp. 275–284). Perth: Murdoch University.

Caldwell, J. T. (1995). *Televisuality: Style, crisis, and authority in american television*. New Brunswick, NJ: Rutgers University Press.

———. (2003). Second shift media aesthetics: Programming, interactivity and user flow. In A. Everett and J. T. Caldwell (Eds.), *New media: Theories and practices of digitextuality* (pp. 127–144). London: Routledge.

———. (2004a). Convergence television: Aggregating form and repurposing content in the culture of conglomeration. In L. Spigel and J. Olsson (Eds.), *Television after television: Essays on a medium in transition* (pp. 41–74). Durham, NC: Duke University Press.

———. (2004b). Industrial geography lessons, socio-professional rituals and the borderlands of production cultures. In N. Couldry and A. McCarthy (Eds.), *MediaSpace: Place, scale, and culture in a media age* (pp. 163–189). New York: Routledge.

———. (2008). *Production culture: Industrial reflexivity and critical practice in film and television*. Durham, NC: Duke University Press.

Carey John and Elton, M. C. J. (2010). *When media are new: Understanding the dynamics of new media adoption and use*. Ann Arbor: University of Michigan Press.

Cialdini, R. B. (2001). Harnessing the science of persuasion. *Harvard Business Review, 79*(9), 72–81.

Coleman, S. (2006). How the other half votes: Big brother viewers and the 2005 general election. *International Journal of Cultural Studies, 9*(4), 457–479.

Corner, J. (1999). British TV Dramadocumentary: Origins and developments. *Why Docudrama*, 35–46.

Couldry, N. (1998). The view from inside the "simulacrum": Visitors' tales from the set of Coronation street. *Leisure Studies, 17*(2), 94–107.

———. (2001). *The place of media power: Pilgrims and witnesses of the media age*. London: Routledge.

———. (2002). Playing for celebrity: Big brother as ritual event. *Television & New Media, 3*(3), 283–293.

———. (2003). *Media rituals*. New York: Routledge.

———. (2004). Liveness, "reality" and the mediated habitus from television to the mobile phone. *The Communication Review, 7*(4), 353–361.

Cubitt, S. (1991). *Timeshift: On video culture*. London: Routledge.

Dayan, D. and Katz, E. (1994). *Media events: The live broadcasting of history*. Cambridge, MA: Harvard University Press.

Doron, G. (1998). The politics of mass communication in Israel. *The Annals of the American Academy of Political and Social Science, 555*(1), 163–179.

Downes, E. J. and McMillan, S. J. (2000). Defining interactivity: A qualitative identification of key dimensions. *New Media & Society, 2*(2), 157–179.

Duffy, B. E., Liss-Mariño, T. and Sender, K. (2011). Reflexivity in television depictions of media industries: Peeking behind the gilt curtain. *Communication, Culture & Critique, 4*(3), 296–313.

Durlak, J. T. (1987). A typology for interactive media. *Communication Yearbook, 10*, 743–757.

Einav, G. (2004). *I want my ITV? Content, demand and social implications of interactive television.* (Unpublished PhD). Columbia University, New York.

Ellis, J. (1982). *Visible fictions: Cinema, television, video*. New York: Routledge.

———. (2000). *Seeing things: Television in the age of uncertainty*. London: I. B. Tauris.

———. (2002). *Seeing things: Television in the age of uncertainty*. London: I. B. Tauris & Co. Ltd.

———. (2004). Television production. In R. C. Allen and A. Hill (Eds.), *The television studies reader* (pp. 275–292). New York: Routledge.

Enli, G. S. (2009). Mass communication tapping into participatory culture: Exploring strictly Come Dancing and Britain's Got Talent. *European Journal of Communication, 24*(4), 481–493.

Enli, G. S. and Ihlebæk, K. A. (2011). "Dancing with the audience": Administrating vote-ins in public and commercial broadcasting. *Media, Culture & Society, 33*(6), 953–962.

Ericson, S. and Riegert, K. (Eds.). (2010). *Media houses: Architecture, media, and the production of centrality*. New York: Peter Lang Publishing.

Evans, W. (2005). Reality programming: Evolutionary models of film and television viewership. In J. D. Anderson and B. F. Anderson (Eds.), *Moving image theory: Ecological considerations* (pp. 200–214). Carbondale: Southern Illinois University Press.

Feuer, J. (1983). The concept of live television: Ontology as ideology. In E. A. Kaplan (Ed.), *Regarding television: Critical approaches—An anthology* (pp. 12–22). Los Angeles: American Film Institute.

Fiske, J. and Hartley, J. (1978). *Reading television*. London: Methuen & co.

Friedberg, A. (1990). A denial of difference: Theories of cinematic identification. In E. A. Kaplan (Ed.), *Psychoanalysis and cinema* (pp. 36–45). New York: Routledge.

Gilder, G. (1994). *Life after television*. New York: W. W. Norton & Co., Inc.

Gill, R (2010) "Life is a pitch": Managing the self in new media work. In Mark Deuze (ed.) *Managing Media Work*. London: Sage.

Gitlin, T. (1993). Bites and blimps: Chunk news, saavy talk and the bifurication of American politics. In P. Dahlgren and C. Sparks (Eds.), *Communication and citizenship: Journalism and the public sphere* (pp. 119–136). London: Routledge.

Goffman, E. (1959). *The presentation of self in everyday life*. Garden City, NY: Doubleday.

———. (1974). *Frame analysis: An essay on the organization of experience*. New York: Harper & Row.

Gray, J. (2010). *Show sold separately: Promos, spoilers, and other media paratexts*. New York: NYU Press.

Greenzweig, E. (2011). Hackers to survivor's production computers indicted. *Walla*, May 11. http://b.walla.co.il/?w=/3050/1822404 (accessed July 24, 2017).

Gripsrud, J. (2010). *Relocating television: Television in the digital context*. London: Routledge.

Hall, S. (1980). Encoding/Decoding. In S. Hall, D. Hobson, A. Lowe, and P. Willis (Eds.), *Culture, media, language: Working papers in cultural studies, 1972–1979* (pp. 122–138). London: Hutchinson.

Handelman, D. (2003). Towards the virtual encounter: Horton's and wohl's' mass communication and para-social interaction.' In E. Katz, J. P. Durham, T. Liebes, and A. Orloff (Eds.), *Canonic texts in media research: Are there any? should there be? how about these?* Cambridge, MA: Polity Press, 137–151

Hartley, J. (2008). *Television truths: Forms of knowledge in popular culture*. Oxford, U.K: Wiley-Blackwell.

———. (2011). *The uses of digital literacy*. Piscataway, NJ: First Transaction Publication.

Havens, T. and Lotz, A. D. (2012). *Understanding media industries*. Oxford: Oxford University Press.

Hawes, W. (2001). *Live television drama, 1946–1951*. Jefferson, NC: McFarland.

Hawkins, G. (2001). The ethics of television. *International Journal of Cultural Studies, 4*(4), 412–426.

Hellman, H. and Keinonen, H, (2013). Semi-commercial or semi-public service? Legitimacy and regulation of commercial television in Finland. *Comunicazioni Sociali, 3*, 44–57.

Herzog, H. (1941). On borrowed experience. an analysis of listening to daytime sketches. *Studies in Philosophy and Social Science, 9*(1), 65–95.

Hesmondhalgh, D. and Baker, S. (2011). Toward a political economy of labor in the media industries. *The Handbook of Political Economy of Communications*, 381–400.

Hill, A. (2002). Big brother: The real audience. *Television & New Media, 3*(3), 323–340.

———. (2005). *Reality TV: Audiences and popular factual television*. London: Routledge.

———. (2016). Push–Pull dynamics: Producer and audience practices for television drama format the bridge. *Television & New Media, 17*(8), 754–768.

Holmes, S. (2004). But this time you choose! Approaching the "Interactive" audience in reality TV. *International Journal of Cultural Studies, 7*(2), 213–231.

———. (2008a). *The quiz show*. Edinburgh: Edinburgh University Press.

———. (2008b). The viewers have . . . taken over the airwaves? Participation, reality TV and approaching the audience-in-the-text. *Screen, 49*(1), 13–31.

———. (2009). "This outlet for criticism could ginger up the whole climate in which television is watched": (Re)viewing *points of view*—audience, access and participation. *Critical Studies in Television: Scholarly Studies in Small Screen Fictions, 4*(1), 4–25.

Horton, D. and Wohl, R. (1956). Mass communication and parasocial interaction: Observation on intimacy at a distance. *Psychiatry, 19*(3), 215–229.

Hutchby, I. (2001). Technologies, texts and affordances. *Sociology, 35*(2), 441–456.

Izikovich, G. (2010). Lyala Bekef: Not afraid to look stupid. *Achbar Haeer*, February 12. http://www.mouse.co.il/CM.articles_item,1025,209,45579,.aspx (in Hebrew) (accessed November 15, 2012).

Janlert, L. E. and Stolterman, E. (2017). The Meaning of Interactivity, Some Proposals for Definitions and Measures. Human-Computer Interaction, 32(3), 103–138.

Jarrett, K. (2008). Interactivity is evil! A critical investigation of web 2.0. *First Monday, 13*(3), 27.11.2010.

Jenkins, H. (2006). *Convergence culture: Where old and new media collide*. New York: New York University Press.

Jensen, J. F. (1998a). Communication research after the mediasaurus? digital convergence, digital divergence. *Nordicom Review, 12*(1), 39–52.

———. (1998b). Interactivity: Tracing a new concept in media and communication studies. *Nordicom Review, 12*(2), 185–204.

Jensen, J. F. and Toscan, C. (Eds.). (1999). *Interactive television: TV of the future or the future of TV?* Aalborg, Denmark: Aalborg University Press.

Jermyn, D. and Holmes, S. (2006). The audience is dead; long live the audience!: Interactivity, "telephilia" and the contemporary television audience. *Critical Studies in Television: Scholarly Studies in Small Screen Fictions, 1*(1), 49–57.

Johnson, D. (2007). Inviting audiences in: The spatial reorganization of production and consumption in "TVIII." *New Review of Film and Television Studies, 5*(1), 61–80.

Katz, E. and Scannell, P. (2009). *The end of television? Its impact on the world (so far)*. Thousand Oaks, CA: Sage Publications, Inc.

Katz, Y. (2012). Media policy in Israel: The new structure of public broadcasting. *Journalism and Mass Communication,* 2(5), 581–596.

Kelleher, T. (2009). Conversational voice, communicated commitment, and public relations outcomes in interactive online communication. *Journal of Communication, 59*(1), 172–188.

Kelly, J. P. (2017). Television by the numbers. *Convergence*, doi:10.1177/1354856517700854.

Keinonen, H. and Klein Shagrir O. (2017). "From public service broadcasting towards soci(et)al television? Broadcasters' perceptions of audiences and interactivity in a multiplatform environment in Finland and Israel," *Nordicom Review, 38*, 65–79.

Kim, P. (2001). New media, old ideas: The organizing ideology of interactive TV. *Journal of Communication Inquiry, 25*(1), 72–88.

Kim, P. and Sawhney, H. (2002). A machine-like new medium - theoretical examination of interactive TV. *Media, Culture & Society, 24*(2), 217–233.

Kiousis, S. (2002). Interactivity: A concept explication. *New Media & Society, 4*(3), 355–383.

Kjus, Y. (2009a). Everyone needs idols: Reality television and transformations in media structure, production and output. *European Journal of Communication, 24*(3), 287–304.

———. (2009b). Fabricating cultural events: The rise of international programme formats in Norwegian television production. *Mediekultur, 25*(47), 92–105.

———. (2009c). Idolizing and monetizing the public: The production of celebrities and fans, representatives and citizens in reality TV. *International Journal of Communication, 3*, 277–300.

———. (2009d). Impact of prestige programs on production practices: The case of crossmedia and audience participation in public service organizations. *Journal of Media Practice, 10*(2 & 3), 167–184.

Klein Shagrir, O. (2015). Unveiling television's apparatus on screen as a 'para-interactive'strategy. *Media, Culture & Society, 37*(5), 737–752.

Klein Shagrir, O. and Keinonen, H. (2014). Public service television in a multi-platform environment: A comparative study in Finland and Israel. *VIEW Journal of European Television History and Culture, 3*(6), 14–23.

Lachman-Meser, D. (1997). The new communications map in israel: Explanation of a process, characteristics, and regret. *Dvarim Achadim, 5*(2), 1–22.

Levine, E. (2008). Distinguishing television: The changing meanings of television liveness. *Media, Culture & Society, 30*(3), 393–409.

Liebes, T. (1988). Cultural differences in the retelling of television fiction. *Critical Studies in Mass Communication, 5*(4), 277–292.

———. (1998). Television's disaster marathons. *Media, Ritual and Identity,*71–84.

Lindlof, T. R. and Taylor, B. C. (2010). *Qualitative communication research methods*. Thousand Oaks, CA: Sage.

Livne, Neri. (2009). *Ha'aretz* weekend magazine. Oct. 2.

Lotman, Y. (1990). *The universe of the mind.* London: IB Tauris & CO.

Lotz, A. D. (2004). Textual (im)possibilities in the US post-network era: Negotiating production and promotion processes on lifetime's any day now. *Critical Studies in Media Communication, 21*(1), 22–43.

———. (2007). *The television will be revolutionized.* New York: NYU Press.

Lunt, P. (2004). Liveness in reality television and factual broadcasting. *The Communication Review, 7*(4), 329–335.

MacCannell, D. (1973). Staged authenticity: Arrangements of social space in tourist settings. *The American Journal of Sociology, 79*(3), 589–603.

Manovich, L. (2001). *The language of new media.* Cambridge, MA: MIT Press.

Marriott, S. (2007). *Live television: Time, space and the broadcast event.* Thousand Oaks, CA: Sage.

Mayer, V. (2016). "The places where audience studies and production studies meet." *Television & New Media, 17*(8), 706–718.

Mayer, V., Banks, M., J., and Caldwell, J. T. (Eds.). (2009). *Production studies: Cultural studies of media industries*. New York: Routledge.

McMillan, S. J. and Hwang, J. S. (2002). Measures of perceived interactivity: An exploration of the role of direction of communication, user control, and time in shaping perceptions of interactivity. *Journal of Advertising, 31*(3), 29–42.

McQuail, D. (2010). *Mass communication theory: An introduction.* London: Sage Publications, Inc.

Meyrowitz, J. (1985). *No sense of place: The impact of electronic media on social behavior.* New York: Oxford University Press.

———. (2003). "Canonic anti-text: Marshall McLuhan's understanding media." In E. Katz, J. P. Durham, T. Liebes and A. Orloff (Eds.), *Canonic texts in media research: Are there any? should there be? how about these?* Cambridge, MA: Polity Press, 191–212.

Moshe, M. (2012). "Media time squeezing the privatization of the media time sphere." *Television & New Media, 13*(1), 68–88.

Murray, S. and Ouellette, L. (Eds.). (2009). *Reality TV: Remaking television culture*. New York: NYU Press.

Nee, R. C. and Dozier, D. M. (2017). Second screen effects: Linking multiscreen media use to television engagement and incidental learning. Convergence, 23(2), 214–226.

Neiger, M. (2012). Cultural oxymora: The Israeli Idol negotiates meanings and readings. *Television & New Media, 13*(6), 535–550.

Nightingale, V. and Dwyer, T. (2006). The audience politics of "enhanced" television formats. *International Journal of Media and Cultural Politics, 2*(1), 25–42.

Olsson, J. and Spigel, L. (2004). *Television after TV: Essays on a medium in transition.* Durham, NC: Duke University Press.

Ortner, S. B. (2010). Access: Reflections on studying up in Hollywood. *Ethnography, 11*(2), 211–233.

Palmer, G. (2002). Big brother: An experiment in governance. *Television & New Media, 3*(3), 295–310.

———. (2006). "Coming together: Thedatingchannel.com and the future of television." *Journal of Media Practice, 7*(2), 109–121.

Peters, J. D. (1999). *Speaking into the air: A history of the idea of communication*. Chicago: University of Chicago Press.

Porat, Yoni. (2009). *Mako Entertainment News*. Dec. 3.

Prensky, M. (2001). Digital natives, digital immigrants part 1. *On the Horizon, 9*(5), 1–6.

Rafaeli, S. (1988). Interactivity: From new media to communication. In R. P. Hawkins, J. M. Wiemann and S. Pingree (Eds.), *Advancing communication science: Merging mass and interpersonal processes* (pp. 110–134). Newbury Park, California: Sage Publications.

Rice, R. E. (1999). Artifacts and paradoxes in new media. *New Media & Society, 1*(1), 24–32.

Roig, A., San Cornelio, G., ArdËvol, E., Alsina, P., and PagËs, R. (2009). "Videogame as media practice: An exploration of the intersections between play and audiovisual culture." *Convergence: The Journal of Research into New Media Technologies, 15*(1), 89–103.

Rosen, J. (2006, June 27). The people formerly known as the audience. *PressThink*. http://archive.pressthink.org/2006/06/27/ppl_frmr.html.

Ross, S. M. (2008). *Beyond the box: Television and the internet*. Malden, MA: Blackwell.

Ryan, M. (1994). Immersion vs. interactivity: Virtual reality and literary theory. *Substance, 28*(2), 110–137.

Scannell, P. (1996). *Radio, television and modern life: A phenomenological approach*. Oxford, UK: Blackwell.

———. (2000). "For-anyone-as-someone structures." *Media, Culture & Society, 22*(1), 5–24.

———. (2002). "Big brother as a television event." *Television & New Media, 3*(3), 271–282.

———. (2003). The brains trust: A historical study of the management of liveness. In S. Cottle (Ed.), *Media organizations and production*. London: Sage.

Sconce, J., Spigel, L., and Olsson, J. (Eds.). (2004). *Television after TV: Essays on a medium in transition*. Durham, NC: Duke University Press.

———. (2004). What if? Charting television's new textual boundaries. In L. Spigel, and J. Olsson (Eds.), *Television after TV: Essays on a medium in transition* (pp. 93–112). Durham, NC: Duke University Press.

Shahaf, S. (2009). *Prime time postzionism—Negotiating Israeliness through global television formats.* (Unpublished PhD). University of Texas, Austin.

Shaked, Raanan. (2009). "My Next Start-Up." *7 Days Magazine, Yedioth Ahronoth*, Jan. 1.

Skeggs, B., Thumim, N., and Wood, H. (2008). "Oh goodness, I am watching reality TV": How methods make class in audience research. *European Journal of Cultural Studies, 11*(1), 5–24.

Skopal, P. (2007). "The adventure continues on DVD": Franchise movies as home video. *Convergence: The International Journal of Research into New Media Technologies, 13*(2), 185–198.

Sørensen, I. E. (2015). Go crowdfund yourself: Some unintended consequences of crowdfunding for documentary film and industry in the U.K. In G. Lovink, N. Tkacz, and P. de Vries, (Eds.) *MoneyLab Reader: An Intervention in Digital Economy*. Amsterdam: Institute for Networked Cultures, pp. 268–280.

———. (2016a). Content in context: The impact of mobile media on the British TV industry. *Convergence,* 1–16, 1354856516681703.

———. (2016b). The revival of live TV: Liveness in a multiplatform context. *Media, Culture & Society, 38*(3), 381–399.

Spector, Dana. (2008). "Behaving Accordingly." *7 Nights Magazine, Yedioth Ahronoth*, Dec. 12.

Steuer, J. (1992). Defining virtual reality: Dimensions determining telepresence. *Journal of Communication, 42*(4), 73–93.

Strange, N. and Bennett, J. (Eds.). (2011). *Television as digital media*. Durham, NC: Duke University Press.

Sundet, V. S. and Ytreberg, E. (2009). Working notions of active audiences: Further research on the active participant in convergent media industries. *Convergence: The International Journal of Research into New Media Technologies, 15*(4), 383–390.

Thompson, J. B. (1990). *Ideology and modern culture*. Stanford, CA: Stanford University Press.

Tincknell, E. and Raghuram, P. (2002). Big Brother: Reconfiguring the "active" audience of cultural studies? *European Journal of Cultural Studies, 5*(2), 199–215.

Tolson, A. (2006). *Media talk: Spoken discourse on TV and radio*. Edinburgh: Edinburgh University Press.

Turner, G. and Tay, J. (Eds.). (2009). *Television studies after TV: Understanding television in the post-broadcast era*. New York: Routledge.

Turner, G. (2010). *Ordinary people and the media: The demotic turn*. London: Sage Publications Ltd.

Turovitz, H. (2010). They are afraid. *Pnay Plus*, July 8, p. 15.

Uricchio, W. (2002). Old media as new media: Television. In D. Harries (Ed.), *The new media book* (pp. 219–230). London: BFI Publishing.

———. (2004). Television's next generation: Technology/interface culture/flow. In L. Spigel and J. Olsson (Eds.), *Television after TV: Essays on a medium in transition* (pp. 163–182). Durham, NC: Duke University Press.

Van de Ven, A. H. and Poole, M. S. (1995). Explaining development and change in organizations. *Academy of Management Review, 20*(3), 510–540.

Van den Bulck, H. and Enli, G. S. (2014). Bye bye "Hello ladies?" In-vision announcers as continuity technique in a European post linear television landscape: The case of Flanders and Norway. *Television & New Media, 15*(5), 453–469.

Van Dijk, J. A. G. M. and De Vos, L. (2001). Searching for the Holy Grail: Images of interactive television. *New Media & Society, 3*(4), 443–465.

Van Zoonen, L. (2004). Imagining the fan democracy. *European Journal of Communication, 19*(1), 39–52.

Wardrip-Fruin, N. and Montfort, N. (Eds.). (2003). *The new media reader*. Cambridge, MA: MIT Press.

Weber, M. S., and Monge, P. R. (2014). Industries in turmoil: Driving transformation during periods of disruption. *Communication Research*, doi: 0093650213514601.

Weick, K. E. and Quinn, R. E. (1999). Organizational change and development. *Annual Review of Psychology, 50*(1), 361–386.

White, M. (2006). Television and internet differences by design: Rendering liveness, presence, and lived space. *Convergence: The International Journal of Research into New Media Technologies, 12*(3), 341–355.

Williams, F., Rice, R. E., and Rogers, E. M. (1988). *Research methods and the new media*. New York: Free Press.

Williams, R. (1974). *Television: Technology and cultural form*. London: Fontana.

Wood, H. (2007). Television is happening: Methodological considerations for capturing digital television reception. *European Journal of Cultural Studies, 10*(4), 485–506.

———. (2008). *Talking with television: Women, talk shows, and modern self-reflexivity*. Urbana: University of Illinois Press.

Wu, G. (1999). Perceived interactivity and attitude toward website. *Proceedings of the Annual Conference of American Academy of Advertising,* Albuquerque, New Mexico.

Yang, F. and Shen, F. (2017). Effects of web interactivity. *Communication Research,* doi:10.1177/0093650217700748

Ytreberg, E. (2009). Extended liveness and eventfulness in multi-platform reality formats. *New Media & Society, 11*(4), 467–485.

Index

About the Author

Oranit Klein Shagrir is a senior lecturer at Hadassah Academic College, Jerusalem, and a course coordinator at the Open University of Israel. Prior to her academic career she worked for the Israeli public television in various production roles.

www.ingramcontent.com/pod-product-compliance
Lightning Source LLC
LaVergne TN
LVHW100353290926
849146LV00001B/4
* 9 7 8 1 4 9 8 5 4 0 8 0 3 *